T0146233

PRIVATE
PRACTICE

PRIVATE PRACTICE

In the Early Twentieth-Century
Medical Office of Dr. Richard Cabot

Christopher Crenner

THE JOHNS HOPKINS UNIVERSITY PRESS
Baltimore & London

© 2005 The Johns Hopkins University Press
All rights reserved. Published 2005
Printed in the United States of America on acid-free paper
2 4 6 8 9 7 5 3 1

The Johns Hopkins University Press
2715 North Charles Street
Baltimore, Maryland 21218-4363
www.press.jhu.edu

Library of Congress Cataloging-in-Publication Data
Crenner, Christopher, 1961–
Private practice: in the early twentieth-century medical office of Dr. Richard Cabot /
Christopher Crenner.
p. cm.
Includes bibliographical references and index.
ISBN 0-8018-8117-X (hardcover : alk. paper)
1. Cabot, Richard C. (Richard Clarke), 1868–1939—Career in medicine. 2. Medi-
cine—Practice—United States—20th century. 3. Physicians (General practice)—
United States—History—19th century. 4. Physician and patient. I. Title.
R729.C74 2005
610.73—dc22   2004022316

A catalog record for this book is available from the British Library.

CONTENTS

PREFACE

History confronts us at times with the strange and the strangely familiar side by side. In examining the records of the Boston medical practice of Richard Clarke Cabot (1868–1939), I sometimes glimpsed a semblance of our present world, but across a gulf of years that made what appears obvious today seem unexpectedly odd. Many elements of Cabot's practice and many aspects of the stories that his patients told him were immediately recognizable: the personal difficulties involved, the engagement of the doctor, and the progression or remission of illnesses over time. Yet Cabot and his patients struggled against illness in a world separated from ours by a century of remarkable change. The contours of Cabot's medical influence and the authority manifest in his clinical work and in his patients' responses differ strikingly from their parallels today.

A photograph of Cabot working at his microscope provides a useful analogy for this balance between the strange and the familiar in my sense of the relationship between this physician and his patients. The picture (see next page) displays a prominent American doctor near the peak of his career around 1900, but it seems at first glance to fit comfortably with the typical iconography of the physician-scientist today. It perhaps calls to mind a similar picture of a doctor standing in front of an illuminated x-ray. The clues to disease lie there before the physician, to be revealed by modern science. Cabot notes down his findings, intent on sorting out the evidence brought to light by his instrument. Like the doctor standing in front of the x-ray, he is employing intricate technology and specialized knowledge to grapple with sickness.

But the resemblance to present circumstances is deceptive. The image of a physician at the microscope at the turn of the century carried markedly

Richard Cabot taking notes on observations through a microscope, ca. 1900. Courtesy of the Harvard University Archives.

different significances. Research, discovery, and daily medical practice were bound up together in this single image in a way that would be impossible to replicate a hundred years later. Gazing through his microscope in 1900, Cabot was pushing against the boundaries of medical knowledge. His first book in 1896 on the diagnostic examination of blood was a foundational text on the subject. Cabot was among the first American physicians to employ the differential staining of white blood cells systematically in clinical practice, work that put him at the forefront of American medical science. Yet the microscope was a common medical tool in Cabot's time; it was being used in doctors' offices and hospitals around the country. It is as though gene splicing today, although still a novel science, were being routinely tried out by doctors to see how it might help their patients. Understood in this context, the picture of Cabot at the microscope evokes a different era.[1]

Looking carefully at the photograph, one notices other differences as well. The modern doctor, seen standing before an x-ray today, would likely be surrounded by clean, fluorescent-lit surfaces, deep inside the sterile halls of a hospital. But Cabot sits in a wicker chair at a writing table, surrounded by bookshelves, lit from the side by an unseen window. There are no laboratory bench, institutional hardware, or hospital personnel. He is in his study at home, perhaps. What is he doing with this cutting-edge medical technology in his study? The picture calls up a vision of the genteel Victorian scientist, and that ideal may have held special sway in an old, established city like the Boston of Cabot's day. But another more fundamental context for the image should also be acknowledged. Cabot in this photograph may well be doing routine medical work in his private medical office, where he typically saw patients at his home in Boston's Back Bay neighborhood. Given what we know about the private medical practices of Cabot and his colleagues, he could be examining a blood sample from a patient waiting in next room. Private practice, at least as represented by privileged urban practitioners like Cabot, often took place in just such settings. These physicians worked in their patients' homes and in offices in their own homes. The hospital was steadily gaining visibility as the new institutional basis of American medicine, especially among the aspiring practitioners who were Cabot's closest colleagues. But private practice remained solidly and routinely domestic in 1900. Cabot is shown at home in this picture, whether he is doing the daily work of medical practice or exploring the potential of a novel medical technique.[2]

Absent from this picture also is the usual white lab coat of the physician-scientist of later decades.[3] Cabot wears a dark suit, starched collar, and tie at his microscope, attire typical not only of his peers in contemporary medical school and hospital photographs but of the contemporary American elite as a class. This was a time when one's place in society still determined in part what kind of doctor one was, more than simply being a doctor determined one's place in society. American medicine in Cabot's day was a rapidly changing profession, with fluid boundaries, but a great number of social strata separated its various practitioners. The Richard Cabot of this photograph might have commanded respect as much by reason of his membership in a socially elite Boston family, the Cabots, as he did by reason of being a physician. But Cabot was a witness to—and sometimes a self-conscious participant in—a large-scale shift in the profession. Within a generation of Cabot's heyday, the physician, donning a white coat, would assume a more impersonal status and authority that seemed independent of the social standing conferred by a family name. Richard Cabot in his formal dress seated at a desk in this Back Bay study would not have been mistaken for just any doctor in 1900. But in 1900, he may not yet have been able to wield the kind of broadly endorsed technical authority that would be assumed by an anonymous colleague thirty years later standing in a white coat in the corridors of a hospital.[4]

This brief exploration of how images of a physician may reflect differences in medical authority should make clear that in examining the authority of an individual physician like Cabot, we are working with a complex amalgam of ideals and expectations, capacities and behaviors. I would like to begin by characterizing more narrowly what I am considering. My concern is with the physician's personal authority in individual exchanges with patients. Such authority was exhibited, for example, in the persuasiveness of a doctor like Cabot working directly in his office with ailing patients and their families. The records of Cabot's private practice considered here document face-to-face encounters with patients who sought the advantages of technical medical expertise in the early twentieth century. The persuasiveness available to Cabot drew in some part from the reputation of the medical profession collectively. So the influence of organized medicine at that time and the growth of the professional power of physicians are relevant to this inquiry. Securing exclusive licensing restrictions on practice, limiting the dispensing of prescription medications, obtaining wider pub-

lic support for elite medical education, and related professional preroga-
tives all contributed to this individual persuasive influence. But I am not
asking about when or how physicians became professionals, or what it
meant for medicine to be a profession, or about the public sources of professional power. The focus of this book is both more commonplace and
more personal, on the individual relationships constructed in private practice between Richard Cabot and his patients, relationships that would
come to support a highly technical and narrowly defined expression of
medical expertise in the twentieth century.[5]

The book is organized topically around the history of Cabot's private
clinic. Chapter 1 offers an overview of the key analytical questions. Chapter
2 sketches the early days of the clinic. Chapters 3 and 4 deal with issues
such as diagnosis and therapy in the context of Cabot's medical practice,
outlining some of the issues that arose among Cabot, his colleagues, and
his patients. Chapters 5 focuses on evolving definitions of neurosis. Chapter 6 discusses the provision of medical care of the dying, examines
Cabot's account of his elder brother's euthanasia, and concludes with a
look at the care that Cabot himself and his wife, Ella Lyman Cabot, received
during their own final illnesses. Finally, chapter 7 surveys the evolution of
medical relationships and the medical marketplace from Cabot's time to
our own. Taken in sequence the chapters give a feel for the life of a single
private practice as it matured, flourished, and declined in step with the life
and career of its sole proprietor.

ACKNOWLEDGMENTS

A book stands on its own merit or falls short to the demerit of its author alone. But that is not the entire story, of course. Many people have helped me improve this book with their advice and support, and I would like to thank them here. One great benefit of the long period that I spent on this work has been the opportunity to draw on the expertise of so many remarkable people along the way, advocates, colleagues, and critics alike.

I take pleasure first in thanking my colleagues in the history of medicine who read and commented on the major sections of this book over its evolution: in chronological order, Barbara Rosenkrantz, Allen Brandt, Charles Rosenberg, and Robert Martensen. In addition, for conversations, support, advice, and consultation that helped me get the work done, from making graphs to picking a publisher to contriving a twenty-minute version, I owe thanks to Robbie Aronowitz, Jeff Baker, Chris Bass, Conevery Bolton Valencius, Jackie Duffin, Hughes Evans, John Eyler, Lara Freidenfelds, Jeremy Greene, Ryan Gregory, Michael Grey, Jennifer Gunn, Joel Howell, Margaret Humphreys, Rob Kaiser, Nick King, Barron Lerner, Ken Ludmerer, Russ Maulitz, Harry Marks, David Meltzer, Jeff Moran, Dwight Oxley, Steve Peitzman, Marty Pernick, Todd Savitt, Ned Shorter, Marilyn and Richard Silverstein, Nancy Tomes, Sarah Tracy, Keith Wailoo, and John Warner. An earlier version of Chapter 3 appeared in the *Bulletin of the History of Medicine* 76: 1 (2002), 30–35, where it benefited from the careful advice of the journal's anonymous reviewer. I would also like to thank the peer-review panel for the National Library of Medicine, who offered many thoughtful comments on a lengthy chapter outline that became the backbone of the book. An anonymous reviewer for the Johns Hopkins University Press provided an attentive reading that went beyond the call of duty, and Jackie Wehmueller at the Press has been truly adept in helping me to pull the whole thing together. I would also like to thank Peter Dreyer for his excellent review of the final manuscript.

This book is about the practice of medicine in an earlier time. But I have drawn unavoidably from my experiences of medicine today. I take the greatest pleasure in thanking the physicians who have shown me how clinical medicine is done well through their attention to its spirit and its craft. As is the sadly common fate of teaching everywhere, the skilled teaching of clinical medicine, although invaluable, is only modestly celebrated. I would like to offer my deepest gratitude to Anita Fines, Robert Levy, Chris Fanta, Marshall Wolf, Bill Taylor, Jan Shorey, David Hughes, Tim Ferris, Steve Gordon, Tom Inui, and Eric Krakauer. Given the nature of medical practice, it is not surprising that several decent, earnest physicians have also provided me with examples of how to fail; and although they were helpful in their own way, I shall not thank them individually.

I received advice on statistical analysis and sampling from Matthew Mayo in the Department of Preventive Medicine at University of Kansas School of Medicine and, in an earlier phase of the project's development, from Joel Perlmann at the School of Education at Harvard University— although they did not review the final outcome and responsibility for the correctness of the analysis rests with me alone.

Richard Hocking, former trustee of the Richard Clarke Cabot Papers, was supportive of this project in its early stages and provided the access to the records that made it possible. Richard J. Wolfe and Elin Wolfe, and subsequently Tom Horrocks and Jack Eckert, in Rare Books and Special Collections at the Francis A. Countway Library of Medicine in Boston were enormously helpful with access to the remarkable resources there. For use of the clinical records of the Massachusetts General Hospital, Dr. John Stoeckle supplied a crucial introduction, and Janet Saunders, Mary Lahey, and later Jeff Mifflin offered generous support and consideration. Rao Korivi and William Doran at Boston City Hospital were similarly important in giving me access to the historical clinical records there. Harley Holden of the Harvard University Archives was a valued advisor and early supporter of my work with the Cabot Papers, while Brian A. Sullivan has provided assistance all along the way.

Jerry Menikoff, Allen Greiner, and Martha Montello, my colleagues at the University of Kansas, offered advice and encouragement at crucial late stages in this work. I am also indebted for assistance with research, bibliography, and copyediting to Sally Heaston, Cherie Kelly, and Renée Sexton. Matthew Scanlon contributed his keen eye and a gifted sense of design in editing and organizing the images for this book.

A National Mellon Fellowship in the Humanities dissertation support grant assisted with some of the early research for this project. The Hartford Foundation provided support at a critical stage of this work through a career development award, and a Publication Grant from the National Library of Medicine through the National Institutes of Health allowed me to carve out the necessary time to complete the final manuscript.

Finally, I thank Jim Crenner for blending pragmatic commentary with inspiration in counseling on the craft of writing.

CHAPTER I

THE AUTHORITY
OF A SCIENTIFIC DOCTOR

W e tend to see the growth of medical authority through the lens of recent events. The past few decades have witnessed ongoing attempts to empower patients with greater autonomy and influence over their medical care, typically in a hospital against a backdrop of dripping plastic tubing and beeping electronic monitors. In this setting, the relationship between patient and physician seems impersonal and steeply hierarchical, and the struggle to empower the patient, envisaged as hedged about and overwhelmed by a tangle of medical machinery, has at times become a struggle against the authority of the individual physician. Scientific tools and techniques have correspondingly been cast as the instruments used to acquire and reinforce medical authority, limiting our perspective on the history of doctor-patient relationships. The technical pursuit of disease with esoteric rationales and shiny instruments seems responsible for a new level of medical power that has deprived patients of their independence and control. Prior accounts of this situation have provided a compelling moral critique of flaws in modern medical care but have had little to say about the development of medical authority.[1]

My examination of the practice of Richard Cabot helped me to expand on these accounts of the growth of medical authority and to define some of their weaknesses. Cabot's colleagues and his peers refined the technical pursuit of disease in the hospital, the operating room, the radiology suite, and the clinical laboratory, but they began this pursuit outside the hospital, where their patients were. The influence of new tools and practices on the medical relationship started in homes and offices. In these settings, physicians were limited in their ability to assume the care of passive, dependent patients. They shared that responsibility equally with patients' family members and other caretakers. Cabot's private practice had an active and independent clientele. His patients were sometimes deferential in their correspondence, but they were equally likely to take issue with the care he recommended. They typically came to the clinic of their own accord and took from it only what they chose. Cabot and his colleagues had to make choices in routine practice about how to apply their new tools and rationales. Which practices would patients accept and value, and how well could physicians justify them? Cabot and his colleagues had to make strategic compromises about the technical care of their patients, compromises that I found neatly documented in the records of Cabot's practice. I initially went to these archives with another set of questions in mind. What I found there, however, led me into the writing of this book. Cabot was especially wary about the difficult choices that he faced in delivering medical care in his private office. In his wariness, he succeeded in making these compromises visible both to posterity, through his meticulous documentation, and also to his patients.[2]

Cabot's records offer insight into a paradox in the development of modern medical relationships. The technical pursuit of disease underwrote a doctor-patient relationship that was, in hindsight at least, increasingly distant and hierarchical. Yet these methods ironically brought patients and physicians closer together. Cabot asked his patients at length about the details of their personal bodily routines and functions, and they responded with thorough accounts of their sexual activity, their bowel habits, or the altered color of their urine, eyes, skin, or tongue. Seeking the sources of disease, Cabot and his colleagues examined their patients minutely, thumping, listening, and peering into the recesses of eyes, ears, rectum, vagina, urethra, or mouth. Such procedures had long been practiced, of course, but the new emphasis on rigor seemed to elevate their importance and extend their use more widely, fostering a new technically mediated physical intimacy between patients and physicians, a feature that seems so characteristic now of our encounters with physicians. The twentieth-century doctor came to cultivate a certain cool interest in everything that was most personal and private about daily life.

Outside the hospital, patients in the early twentieth century did not, however, submit themselves to such technical control. Rather, they delegated and channeled their own interests through their doctor, seeking to get the best of the treatment available. Cabot is a particularly valuable witness to the changes taking place in medicine in this era, because he sampled these technical powers but quickly became aware of their limitations.[3]

Home, Office, and Hospital

Richard Cabot's individual office practice provides an opportunity to examine the vital medical territory that lay between the home and the hospital. Later in the twentieth century, the influence and control of individual physicians over their hospitalized patients was profound, inspiring a mid-century sociologist to name the hospital of his day as one of the "total institutions" that had the capacity to structure all aspects of the daily existence of those who found themselves inside it.[4] Even by 1902, when a physician at Massachusetts General Hospital wrote on a hospital chart that a patient should receive "meats only one a day," and "Rx: Pot[assium] iodide," the note likely documented an event as much as a recommendation. Professional nurses in the hospital helped to realize this medical control.[5] To reap the promises of technical medical practice, physicians early in the twentieth century expanded and reshaped a range of institutions, starting with the hospital. By the mid twentieth century, the system of hospitals, medical schools, professional specialty societies, and research institutes had helped to consolidate an unprecedented degree of professional influence for physicians. Under the aegis of the scientific control of disease, doctors emerged with a persuasive authority as the proprietors of an immense medical system. But the hospital lay at the distant end of a long journey from the domestic contexts of nineteenth-century medical practice. The medical office was often the patient's first step out of the home.[6]

The medical office played a vital role in constituting and organizing American private practice.[7] Among a small but growing group of patients who sought hospital care in the early twentieth century, many came by way of the medical office. A private office like Cabot's thus occupied a crucial transitional site in the public's encounter with modern medical authority: halfway between convalescence in the nineteenth-century home and technical scrutiny in the twentieth-century hospital. As American physicians in the middle of the century celebrated the success of their grander institutions, they tended also to claim that their newfound privileges depended on small-scale personal commitments. They often justified their individual medical authority in terms of their relationships with their patients, typically in the office, in private practices.[8] The medical office offered a serviceable way station between the patient's home and a growing system of self-contained medical institutions. In the transitional setting of the doctor's office, patients and physicians sought agreement on how to manage the alluring but often burdensome resources of technical medicine.

Applying expertise in medical practice differed from demonstrating it in face-to-face encounters with patients. The challenges imposed by this difference were particularly acute in the domestic settings of early twentieth-century medicine, as I hope to make clear in considering Cabot's practice. Applying and showing expertise might be mutually supportive or they might conflict. The distraction of being observed by a patient, for example, could bring about a slip in concentration, while deep absorption in the task at hand might lead to a neglect of those who were watching. Physicians in the twentieth century have been noted to experience especially affecting differences between the "backstage" and the "onstage" performance of their craft.[9] One of Cabot's patients remembered decades later how she had sat in her parlor at home with Cabot while his brother, the surgeon Hugh Cabot, removed her son's appendix in the next room. Hugh then came in bearing the offending appendix on a napkin to show her.[10] Hugh Cabot was crossing a critical boundary when he moved away from the backstage position at the operating table to come onstage to discuss outcomes, carrying evidence of success. Skill and knowledge were critical in cutting out a diseased appendix in 1900, but how this relatively novel procedure and its results were described, explained, justified, promoted, or excused was just as important, as Hugh seems to have realized.

Meanwhile the application of expertise too became more challenging for physicians like the Cabot brothers, who were increasingly devoted to exacting technical practice. Even as public appreciation of medical abilities grew, the task of developing and deploying these skills and techniques became weightier. Technical skills like aseptic surgery required extensive training, refinement, and practice. The diagnosis and treatment of diseases involved ever more tools, more technical data, and more manipulations. In addition more medical activities took place at a distance from the immediate personal contacts between physicians and their patients and families, in the medical laboratory, the surgical suite, and the autopsy room. These offstage processes became both increasingly influential and increasingly difficult to describe and explain, as I hope to show in the details of Cabot's practice. The cultivation of private practice was a basic goal for American doctors, even among the academically inclined physicians in Cabot's circles. It required successful management of this slim margin between appearances and practices within a watchful domestic setting. Physicians had long relied on the willingness of their patients to suspend their skepticism, at least in face-to-face encounters, but their chance of success in getting them to do so improved considerably in the early twentieth century. Physicians came to rely on the public recognition that medical work had a legitimate complexity, in the apt phrase of the sociologist Paul Starr. But this recognition developed slowly, in the context of negotiations with individual patients and their families. There was a legitimate complexity to the story of these background negotiations too.[11]

Sicker and more dependent patients may simply have acquiesced to the abstract manipulations of disease management. The mysteries of esoteric technical method in the right context did, of course, enhance authority and control. Expanding space and resources in the hospital meant that twentieth-century physicians would not be operating in the parlor for long. In the hospital, the surgeon would come under the scrutiny of patients and their kin in spaces that were organized according to medical needs. Public performances of expertise would be brief and neatly orchestrated, creating the basis for a fundamental change in individual medical influence.[12] Yet even the elite academic physicians of Cabot's day still did most of their work in the office or in the patient's home. Arranging medical care in these settings required negotiation and persuasion. Outside of the hospital, professional nurses played a minor role, because it usually remained up to patients and family members to carry out a physician's recommendations—or not. Hospitals would soon become the dominant workplace for American physicians, but at the turn of the century, they had not yet managed to attract the bulk of individual private practice.[13] In the context of the home and the medical office, physicians marshaled their influence carefully and tried to manage a technically sophisticated medicine whose workings were often obscure to patient and family.

In the hospital, the ambiguities of technical practice were still a challenge to physicians, but they remained largely invisible to patients. In 1897, Cabot was attending to a very sick woman, whom he thought had diphtheria, at Massachusetts General Hospital. The new treatment of diph-theria with an antitoxin, developed only in the previous year, was already seeing use at the hospital and might offer a cure. But a bacteriological test for diphtheria by the hospital laboratory returned a negative result, sug-gesting that the patient's condition was not diphtheria. The laboratory results showed instead the presence of a different bacterium in the throat, one that would not respond to the new antitoxin, so it was not used.

Although the antitoxin was known to have dangerous side effects, fail-ure to use it could also prove deadly. On the same hospital wards a month earlier, a different patient with a negative test for diphtheria had not been given antitoxin. Cabot delayed the treatment. Four days later, the chart reported "test ordered again by Dr. Cabot." He remained suspicious that diphtheria was actually the cause. Two days more passed. The second labo-ratory report returned, this time showing "numer[ous] colon[ies] of" diph-theria bacteria. A handwritten note at the bottom of the laboratory report indicated, however, that the "report reached house officers['] hands after death of patient."[14] The patient had died from diphtheria without receiving antitoxin. It is likely that only Cabot and the young physicians working with him ever understood the implications. What part of these complexi-ties and ambiguities in technical practice would become evident to patients outside of the secure setting of the hospital, and how would physicians respond?

The Dilemmas of Scientific Practice

Twentieth-century scientific medicine exercised its persuasive influence in part through a growing set of new tools and techniques.[15] Expansive new technologies, from x-ray photography and aseptic surgery to diagnostic microbiology and diphtheria antitoxin, had both excited the public interest and invested physicians with a growing confidence in their ability to con-trol human disease.[16] In his private office, Cabot progressed in two decades of practice from simple reliance on a stethoscope and chemical urinalysis to the use of serology, intravenous pharmacotherapy, surgical biopsies, and barium-contrast radiography. Technical demonstrations of medicine's power might bolster the trust and confidence of patients. But these utilities of the new medicine might equally frustrate people who felt no better for all its promise, as was the case with one such patient who wrote to Cabot in

1916 lamenting, after her encounters with various surgeries, x-rays, and chemical medications: "I can't believe in this time of such wonderful things as are being accomplished in medical science that there is no help for me."[17] We should be wary of the claim that the persuasive power of physicians resided simply or necessarily in their techniques. The image of Cabot at his microscope in the Preface is a reminder about the capacity of tools and practices to confer different kinds of authority in different contexts. The routine use of new techniques in medicine posed particular challenges to their enthusiasts in the early twentieth century, Cabot prominently among them.

Technical medical practice at the turn of the century left physicians like Cabot facing several related dilemmas. He and his colleagues often found physical abnormalities through blood counts, surgical biopsies, and x-rays that related only abstractly to how their patients felt. Conversely, lengthy diagnostic evaluations also on occasion failed to reveal any source for a patient's discomfort. Was the patient sick if no disease or aberration turned up after much testing and evaluation? Treatment posed similar challenges in routine practice. By 1900, therapeutic strategies that aimed at influencing the biology of disease seemed increasingly likely to contradict a patient's experience of treatment. Cabot wrote to a colleague in 1912 that he could slow the decline in his patient's blood counts "by giving Fowler's Solution [a solution of potassium arsenite] beginning with two drops after each meal and increasing up to the limit of toleration, that is until nausea and diarrhea . . . are produced."[18] The nausea counted only as an incidental side effect that limited treatment, while the sought-after therapeutic effect on the blood count remained imperceptible, at least to the patient. It was in fact a negative result that was sought: to slow a decline in the blood counts. Treatments that aimed to impede the progress of disease, as judged by independent laboratory criteria, might make people feel worse, as an oddly anticipated but unintended consequence of altering the disease.

Scholars of present-day medicine have noted that the modern physician's authority seems to accommodate a potentially disruptive disparity between biomedical definitions of disease and the personal experiences of illness. Yet we have not examined at any length the historical context in which this tension, and the authority to sustain it, emerged.[19] In the early decades of the twentieth century, this disparity became increasingly influential in the daily activities of physicians like Cabot. Awareness of the ten-

Volume 7 of thirty-six bound volumes of Richard Cabot's patient records, open to a page of clinical notes on a patient's visit, with a letter from the patient. Courtesy of the Harvard University Archives.

sion between the perspectives of physician and patient was not new, of course, echoing as it did long-established concern in medicine about the gap between the problems of individual sick people and the abstract patterns of disease. The ancient Hippocratic writings advised the doctor, for example, to watch critically for differences between what they knew from the study of disease and what their patients told them about their conditions.[20] Yet to the physicians of Cabot's day, this distinction between what they observed and what their patients described carried a more pragmatic weight in daily professional life.[21] The technical pursuit of biological disease in the early twentieth century put a novel force behind the traditional hesitation to generalize on the care of individual patients. Certainly, there would be flexibility in using and interpreting the new instruments of scientific medicine. But what limits would physicians impose on themselves and their colleagues in expressing this flexibility? A growing array of technical practices was gradually making disease a more palpable, and potentially eradicable, presence in the clinic—even as these same techniques reinforced the view that the evidence of disease might routinely contradict the patient's experience of illness and healing.

The evidence of this tension between illness and disease lies buried in the medical records of the day, emerging only occasionally in professional comments and open debate. It was during Cabot's career that American physicians first cited a formal distinction between disease as objectively defined by the physician and illness as subjectively experienced by the patient.[22] Yet even as American physicians began to explore this formal distinction between illness and disease, they blurred it into other, safer, better-established dichotomies. Designating illness as the realm of the patient's experience risked a potentially hazardous split. By formally accepting illness as the experience of disease, physicians risked emphasizing the patient's role as the arbiter of the ultimate value of medical care. What if the lab test looked better and the patient felt worse? Rather than debate the problem, physicians of Cabot's time tended instead to corral the issues into a parallel set of debates about medical judgment, emphasizing the differences and tensions among physicians, rather than those between physician and patient. Doctors of the early twentieth century talked relatively little about the differences between illness and disease.

Two other widely considered dichotomies in medicine attracted the main attention of the day. Cabot's peers discussed the distinction between "organic" and "functional" disease at length. And they similarly debated the proper balance between the art and the science of medicine. These two commonly debated—and still potent—distinctions reflected some of the same anxieties about technical, disease-oriented approaches to practice that were raised in considering disease and illness. But the familiar dichotomies between science and art, or between the functional and the structural-organic, kept the issue more safely inside the boundaries of medical definition and control. The patient's perspective remained tangential to the consideration of these more commonly debated dichotomies.

In defining functional disease, for example, physicians contrasted it to organic disease. Organic, or structural, disease left discrete, stable, and detectable marks on the sick person's body—the physical substrates that were made clear by technical methods of measuring, manipulating, removing, and displaying disease. A functional disease, in contrast, was changeable and contingent, only subjectively defined. But it remained within the purview of the physician's judgment.[23] Functional diseases could be transient and variable in different individuals, but the attentive physician who interpreted the signs correctly could still spot and manage these derangements. The doctor who knew the patient's usual state would recognize subtle deviations and learn how to respond in a way that was tailored to the individual constitution and context. The growing power of bacteriology, surgery, and radiology helped to define structural-organic diseases, but equally, by exclusion, defined the functional. This technical power did not deprive physicians of their capacity to recognize functional problems in their patients, as long as the practitioners attended to the distinction. Categorizing diseases into organic and functional permitted a crucial role for the physician's judgment, independent of the patient's subjective experiences. This distinction between functional and structural posed no basic obstacle to the exercise of medical authority, but rather supported individual judgment about the appropriate use of technical methods and their necessary limitations.[24]

Debate over the art and the science of medicine similarly avoided the disruptive implications of a difference between the experience of illness and the treatment of disease. Considerable speculation and debate surrounded these distinctions of disease as functional or organic, and of medical practice as art or science. Yet both distinctions avoided direct challenge to the physician's ability to settle questions about medical knowledge and practice. Discussion of the art and the science of medicine, so popular in Richard Cabot's day, left both these aspects of medicine well inside the scope of a doctor's authority—although different essayists of the day happily argued the relative merits of art and science. Medical science provided an objective vision of natural disease, while the physician's art translated this vision into humane application for the individual patient. Physicians remained the arbiters of the boundaries of subjectivity for their patients and their practices. The phrase "the art of medicine" captured in a single, attractive term the physician's ability to minister compassionately to the difficulties of sickness. Physicians put the power of medical science into use through "the devotion to the relief of suffering, the readiness to help, the sympathy and the kindness of heart," as one leading physician of the day proudly explained in a ceremonial address on the art of medicine.[25] The potential rift created by subjective judgments about sickness lay safely inside the ambit of medicine's vision and control.

The distinction between illness and disease, in contrast, highlighted a fundamental break, since it emphasized the divergent perspectives of physician and patient. In distinguishing illness as the patient's domain, physicians acknowledged their inability to control certain essential evidences about practice. By attending to illness, physicians might confront the limits of an exacting disease model of medical care. Serious consideration of the patient's subjective judgment also prompted a comparison to the subjective judgments of the physician.[26] The physician's judgment and opinion might seem analogous in important ways to the patient's. Medicine would then become no longer a matter only of the expert management of disease, but would include the sick person's judgments about care. The debate in the American medical literature over the tensions between illness and disease remained subdued. Meanwhile, the necessary compromises were being hammered out behind closed doors in clinics, offices, and homes around the country.

Cabot as a Scientific Practitioner

Cabot's clinical records made these tensions between objective disease and subjective illness evident in routine medical practice, in part because Cabot was himself concerned about their implications. As both a convert to scientific medicine and a prominent critic of its weaknesses, he was drawn to the kind of dilemmas that emerged in the routine application of technical methods. Trained in the premier institutions of his day, Cabot aspired to be a leader in the academic medical world. After graduating from Harvard Medical School in 1892, he secured a coveted spot as a house officer at the Massachusetts General Hospital on the East Medical Service in 1894, moving from there to become the director of the hospital's Out-Patient Department and eventually the chief of the prestigious West Medical Service.[27] Cabot staked his early and considerable success in medicine on his expertise in specialized microscopic and serological analysis of the blood. His research on the interpretation of blood tests led to a first book in 1896, *A Guide to the Clinical Examination of the Blood for Diagnostic Purposes,* which outlined the practical, clinical value of blood analysis.[28] Shortly after his book appeared, he volunteered to sail on the SS *Bay State* to assist with transporting sick soldiers back from the Spanish-American War. The returning men had contracted a variety of diseases, such as typhoid and malaria, which Cabot, using his new diagnostic techniques, could identify more readily and with greater confidence than other doctors. His technical facility in diagnosing the soldiers gave him a considerable boost. "No one else knew anything," he recalled later; "I was king."[29] He manifested a similar enthusiasm for the power of technical methods throughout his career. The records from his private medical office demonstrated the steadily expanding application of these methods to daily practice.

While establishing a markedly successful early career in medicine, Cabot preserved a role as a skeptical observer of scientific medicine and of general medical practice. His most vital years in practice—between 1900 and 1915—coincided with his rise through the medical ranks in Boston, one of a handful of important cities for academic medicine in the United States at that time. But all through this period, Cabot built and nurtured a reputation as a critic and gadfly to the medical profession at large. He cultivated the role of participant-observer, situated outside the conventional boundaries of his profession but deeply involved in its activities. He criticized zealous, up-to-date practitioners for relying too heavily on specialized laboratory techniques, but just as readily criticized the old-fashioned doctors for neglecting them.[30] He was a controversial and sometimes unpopular figure among his medical colleagues. Even his prominently featured obituary in the *New England Journal of Medicine,* while observing the expected pieties for the occasion, could not avoid noting Cabot's "militant bluntness and precipitate desire to correct the errors of the practice of medicine."[31] His passion for pointing out errors did, however, earn him a lasting and well-deserved reputation as a keen commentator on the medical scene. He is still widely credited a century later for chiding the American medical profession about the covert use of placebo therapies, and for denouncing the unwillingness of his colleagues to share with their patients the diagnosis of a life-threatening disease.[32]

Cabot's attitude toward his own provisional self-identity as a physician was neatly demonstrated in two private references that he made, bracket-

Cabot's report on his analysis of a sample of blood from a patient at Massachusetts General Hospital in 1898, showing a differential count of the white blood cells, a technique that he helped to pioneer. Massachusetts General Hospital, Patient Records, vol. 501, East Medical Service, tipped in p. 124. Courtesy of the Massachusetts General Hospital, Archives and Special Collections.

ing his medical career at its beginning and end. Writing to his future wife, Ella Lyman, in 1893, just as he was beginning his career, he mused that he would "practice medicine till 45 or till I can live on my income, then I will cultivate other fields."[33] By his late forties, Cabot had, in fact, stepped back from his medical career. After his return from World War I, where he served as chief of the medical staff of a base hospital for the American Expeditionary Force in France, he quickly scaled back his private practice, seeing just 229 patients in all of 1919 and only 40 in 1920. Although he continued to keep some scant practice records on patients into 1926, in 1921, he withdrew from his position as chief of the medical service at Massachusetts General Hospital and professor at Harvard Medical School, accepting a post as professor of social ethics at Harvard College on the main university campus across the Charles River in Cambridge. He had, as he had predicted, turned to cultivate other fields. In later reminiscences, Cabot dated his divergence from medicine earlier, to 1912, when he found himself passed over for an anticipated promotion to the Jackson Professorship of Medicine at Harvard Medical School.[34] Cabot's public good sportsmanship on this occasion drew general praise; but his later private notes suggest a very real disappointment, obliging him to confront his self-imposed status as an outsider in medicine.[35] In reflecting at the end of his life on the derailment of his traditional medical career, Cabot again turned to his differences with the profession. His troubles, he mused in a private note, began with his resistance to a basic dogma in medicine: "I was not interested in Disease, [I was] interested in spotting it and getting people over it. The great men in medicine are interested in Disease. I was happier teaching ethics."[36] Cabot frequently gave the sense that he was in medicine but not "of it," able neither to escape the powerful attractions of the disease model nor to deny its profound limitations.

As the discussion so far suggests, this book will develop a distinct portrayal of Richard Cabot, the physician. Yet it is not a book about the character, personality, or life of Richard Cabot, except in a tangential sense. My intent has been to conscript this physician into an investigation of how technical medical practices engaged and also frustrated the enthusiasm of his colleagues and his patients, and how this tension helped to shape individual medical influence. I have not sought to understand Cabot better by examining his medical practices; rather, I have examined his life and work to the degree that they have helped to interpret his practices. He was an unusual and complicated man, by the accounts of many who knew him, and his struggles with the nature and significance of medical authority perhaps reflected certain wrenching personal experiences early in his career. The reader interested to fathom him deeper as a doctor and a person will find much to consider in the sixth chapter of this book. At a crucial moment when he was starting his medical career, Cabot went through a devastating and surely transforming event. Just as he was finishing his first year in medical practice, he accepted the responsibility of becoming the doctor of his dying elder brother, Ted Cabot. Richard Cabot became convinced that he had to act to end his brother's life to avoid a protracted and agonizing death. The fact of Cabot's direct involvement in his brother's death subsequently became a carefully kept secret. He never made it public or discussed it with his family and avoided stating more than the vaguest general views on medical measures taken to end the life of a patient.[37] So strongly did these memories weigh on Cabot, however, that he made special preparations near the end of his life to set down a detailed account for posterity, long hidden among his personal papers. Meeting with a special confidante, he gave a transcribed interview reviewing the extraordinary circumstances leading up to what became a mysterious family tragedy.[38]

One does not need to delve deeply into the question of the personality of Richard Cabot to see a connection between this momentous event in his early life as a physician and the issues in medicine that later drew his attention and critique.[39] What were the proper limits of the physician's influence and control over patients? A growing technical mastery of bodily disease vested daunting responsibilities in physicians. Medicine offered the power to illuminate and influence matters crucial to patients, even matters of life and death. How would physicians define and deploy this growing power? And how could they make good on the responsibilities that it implied? For Cabot, and for many of his peers, these questions had particular relevance to the relationships that they formed with patients around the exacting technical pursuit of disease. What did it mean to give a placebo medication to a patient in the face of a growing specificity in treatment and an expanding ability to monitor treatment objectively? What obligations did physicians have to disclose the evidence of a life-threatening disease, especially when this evidence derived from independent sources of information like an x-ray or a surgical biopsy? Cabot's pursuit of these dilemmas may have made him sensitive to the quandaries that accompanied the integration of technical, disease-oriented methods into routine practice. His involvement in his brother Ted's death early in his medical career squarely confronted him with fundamental questions about what it meant to be a doctor.

Cabot's Office Practice

To pursue the analyses in this book, I have used an extensive, previously unopened collection of records from Cabot's office practice. Between 1897 and 1926, Cabot built up a private medical practice in central Boston that at its height drew in nearly a thousand patients each year. Over the course of his long career, Cabot set down and archived meticulous records documenting his successful practice. Although his was a small institution, it was neatly situated to provide technical medical services to people who sought them but often accepted them only provisionally. Cabot would eventually fill thirty-six thick, blank folio volumes with detailed notes.[40] His records also collected the extensive correspondence of a large network of colleagues, general physicians, specialists, and other practitioners, who shared patients with Cabot, discussing their care, and providing surgical, laboratory, radiological, and other therapeutic and evaluative services. Most intriguingly, these records preserve the personal letters of the many patients who wrote to Dr. Cabot over the years. People set forth often lengthy accounts of their illnesses and their quest for assistance. In thousands of letters, they interpreted back to their doctor the significance, the advantages, and the shortcomings of his and his colleagues' practices. These detailed records permitted me to explore the provision of a technically complex medicine to the clientele of a single general medical clinic in the early twentieth century.

These records are part of a larger archive of Cabot's personal papers. Indeed, the existence of the papers testifies to Cabot's frank desire to leave behind the evidence of his career for his successors to learn from. The entire collection was already being organized and worked upon during Cabot's lifetime by a potential biographer and close associate, Ada McCormick, who eventually abandoned her biography in the later stages of writing.[41] The private office records, however, also constitute a kind of institutional archive of his practice. They were as much the documentation of the operation of the office as they were his personal papers, and I have read them in the hope of adding to our general understanding of this fascinating institution. Cabot was the sole physician in this practice, the owner and the direc-

tor of his clinic. But the papers clearly represent the products of a complicated operation that extended far beyond Cabot's oversight or control. The
records disclose the evidence of the joint efforts of secretaries, collaborating physicians, laboratory technicians, patients, and their other doctors,
caregivers, and family members who came to the office. They show the
marks of many hands and are interpretable as the artifacts of complex
social events. They record not only the data of a personal, psychological,
and intellectual life but the activities of a private medical practice, dutifully
filed away by its principal architect.

Clearly, Richard Cabot cannot serve in this analysis as a typical or representative physician of his day. He was a deeply committed medical practitioner and a spirited critic of certain aspects of medicine. He was also by
all testimony neither your average doctor nor your average man. Three
additional lines of inquiry have served to balance the narrow focus here on
Cabot's single medical practice and to place the extensive records of his
practice in a larger interpretive context. The medical literature of the early
twentieth century is the most valuable resource. Cabot held sometimes
idiosyncratic views about medicine, but fortunately he was confident in airing them widely in the medical and popular press. It is easy in many
instances to compare Cabot's views on medical practice with those of his
contemporaries and to measure the response of his colleagues in their
published reports to the controversies that he stirred up. In addition, Cabot
wrote extensively about his ideals for routine practice, making it possible to
compare what he did in the clinic with what he wanted to be doing, and
with what his colleagues stated that they should be doing, with often illuminating results. Using the conventional medical literature of the early
twentieth century and Cabot's extensive publications, we can situate him
informatively among his peers and colleagues.

More important, this successful, established private medical practice
meshed smoothly with others, exchanging patients with competing and
collaborating colleagues and clinics. Evidence concerning the network of
professional relationships and contacts that supported Cabot's medical
office provides a crucial resource for situating the functions of this one
clinic. For daily operation, Cabot's clinic was required to fit into a well-
established system of private medical enterprise in early twentieth-century
Boston. He exchanged patients, favors, bills, payments, diagnostic data,
recommendations, opinions, and even blood samples with a large group of

other practitioners, some medical and others not. The careful records of these exchanges in the archives make it possible to see the workings of the whole interrelated market for medical services, at least from one angle. This market constrained Cabot's clinic by its need to function cooperatively, leveling out some of its idiosyncrasies through the exigencies of normal business.

Finally, Cabot's patients expressed their own views about his clinic and its services in letters that were collected and preserved among these records, providing the final context in which to situate this analysis of a single practitioner. The patients' statements about Cabot, his colleagues, and the routine workings of their private practices provide a valuable perspective on the nature of medicine in a single clinic, offering a privileged view that is otherwise very difficult to achieve.

The Patients' View

A new historical attention to the individual experience of illness has in part inspired the approach here, although this book does not attempt to interpret the private experiences of illness and medical care.[42] The records of this clinic instead direct attention to a more specific question about the individual response to medicine in the early twentieth century. The letters from many hundreds of patients who used this clinic offer very selective insights to the reception of technical medical practices. People who wrote to Cabot demonstrate little intent to disclose their true, intimate thoughts about their illness. They wrote for help and advice. Yet their letters disclose much about how they engaged and negotiated with their doctor over care. Cabot's patients generally addressed their letters to him after a recent visit to the office, and sometimes overtly at his prompting—although others wrote unheralded to seek his advice or to arrange a possible consultation.[43] An occasional letter from a spouse or sibling or another physician shows how cautiously patients sometimes constructed their accounts for their doctor. The letters of these patients offered a carefully mediated expression of their interests and predicaments. One great value of these records is the ability to compare the patients' observations on medical service with the specific services that Cabot provided and to follow the exchanges between doctor and patient over time. In their letters, people offered perceptive, sometimes heartfelt, interpretations of medical service. I have used this rich archival collection to examine how a powerful and abstract scientific medicine engaged the interests, interpretations, and cooperation of patients as consumers of their physician's expertise and services.

The sharp contrasts between the patients' accounts and the accounts in Cabot's medical charts provide insight into the double-edged nature of scientific practice, as will become evident especially in chapters 3 and 4. Cabot's conventionally formatted medical records, similar to many preserved from the time, portray a medical authority that derived from an exclusive and privileged knowledge of disease. Medical language and idiom in the chart represent patients as the passive bearers of their diseases and as the recipients of tests and treatments that they barely comprehended or controlled. Medical technique seemed to translate their confusing distress into signs of hidden pathology, almost independently of their own accounts and interpretations. In the standard format of the medical record, Cabot and his colleagues represented diagnosis as a means to penetrate the superficial appearances of a patient's troubles to the deeper roots of disease; and they characterized the effects of their treatments as evident sometimes only through special technical monitoring of disease. Professional documentation in the clinic portrayed the patient as someone who was subjected to medical manipulation without an active role in evaluating or mediating it. But patients recorded a different perspective.

In their letters, patients conveyed a sense of medical authority to Cabot that was negotiable and derived in part from the attractions of a set of powerful, and potentially flexible, services. They wrote to Cabot describing how they had come to use his various resources, attempting to adapt them to their medical problems. Some patients wrote simply to endorse the particulars of Cabot's care. Others, however, spoke of how they had needed to halve a prescribed dose of medicine, to omit it, or to take another remedy of their own contrivance to counteract its harsh effects. They sought to modify Cabot's diagnoses, to conceal a diagnosis from a relative, or to replace it with one more seemingly favorable or appropriate to their problems. They wrote seeking treatments and tests, sometimes ones that he had already declined to give; and sometimes they showed themselves ready to go elsewhere to get what they wanted.[44] People clearly on occasion recognized the differences between what their doctor saw in their problems and what they themselves experienced. Their responses generally reflect recognition that there were legitimate differences of perspective. Although Cabot's patients occasionally tried to trump him, they generally hoped to gain the benefit of his perspective and the advantage of his technical expertise.

Cabot's patients often seem to have recognized the novel tensions that were being created through the wider use of objective tools and techniques for handling disease. They asked about the laboratory evidence of their condition, which they sometimes tested against their own perceptions. A blood test might indicate improving anemia, but what did an improvement in anemia feel like? A treatment made one feel better—but did the serological tests confirm this improvement as well?[45] We might assume that the physician's privileged knowledge of disease would overrule the views of patients who were rendered passive by illness and technical mystery. Yet in order to gather this knowledge and to put it to work, physicians had to recruit patients into a taxing technical process that required extended cooperation and engagement. The records of Cabot's practice preserve the daily accounts of many people who raised questions about the discrepancies that they saw in their physician's methods. These patients actively sought out, paid for, judged, engaged, and declined the technical expertise of Cabot and other doctors. They proved alert on occasion both to the particular attractions of technical medical care and to its capacity to undermine their expectations. Cabot diligently archived many variations on this problem in his notes and in the letters that he elicited from his patients and collected. Some people proved puzzled or merely resigned when gaps arose between their expectations and their doctor's care. Others simply moved on. But some tried to reinterpret their interests and needs to make use of the novel resources commanded by their physician. This book explores these negotiations.

Active Patients

Prompted by such testimony, we might ask whether Cabot, or the structure of his practice, in some way encouraged patients to assume the guise of active and critical consumers of medical services. By a certain point in his career, Cabot's public status as a dissenter in his contemporary medical world inspired a few people to side overtly with him against the conventional medicine of his day—although the best evidence of this dynamic came from the letters of people who were not his patients, but perhaps better described as his fans.[46] Cabot's growing public reputation as a critic may have inspired some part of the response of his patients. But in truth, the orientation of his patients was hardly unique. Roy and Dorothy Porter have emphasized the need to situate the medical practices of the eighteenth century within a broad and often raucous marketplace for health services.[47] Cabot's colleagues and peers were trying, with growing success, to escape from an unconstrained market for their services and to distinguish themselves as fundamentally different from other competitive practitioners of care for the sick. But their efforts had not yet secured many of the safeguards that they later would. Indeed, many of the features of Cabot's practice that would seem likely to have encouraged an active role among his patients were found in the practices of his peers as well. People came to this office typically under their own power and on their own initiative, although family members were also important in this process. Patients paid Cabot and other physicians directly for the services that they obtained. And an inability to get what they wanted sometimes prompted them simply to sample more widely from a growing range of services elsewhere. The uncertainties and incapacities of illness and the daunting complexity of newer technical services may, of course, have blunted the attractions of the role of active consumer and discerning purchaser of medical care. Cabot's patients expressed their own ambivalence about this role at times, suggesting a willingness to commit to an exclusive dependence on Cabot. "I eagerly wait your prescribed regimen which I suppose you will plan for me. I shall try to be faithful in carrying it out," one woman wrote—although she had arrived at Cabot's office preceded by a letter from a prior physician who complained that she had ignored all previous recommendations and skipped out on further appointments.[48] There were elements of Cabot's practice that inspired his patients to greater independence in their quest for medical services, but they were not unique and they surely did not create this activism.

Cabot's records serve also as a reminder of the fact that people who get sick do not necessarily act as neutral informants about their conditions or as passive bearers of their diseases. People instead interpreted and managed their illnesses actively even as they experienced them. The standard rhetoric of medical practice that preceded and survived the early twentieth century obscures this fact, in part, by representing patients as naïve participants in the process of disease identification and treatment. "People usually think in terms of symptoms rather than diseases," an article in a medical journal asserted without qualification in 2001, for example.[49] This observation is a convention of medical reasoning, already well established in a comparable form by Cabot's day. Doctors will find in symptoms the clues to disease and the guides to treatment—but this heuristic will be unavailable to patients. The natural reflexivity of a sick person's experience, moving from the experience to its interpretation and back, has thus been denied within medicine's standard clinical rationales.

But Cabot's patients, at least in addressing their physician, tended to talk about their symptoms most when they were talking about their diseases, either putative or known, mentioning a "cough" only in regard to "an accumulation . . . on the tubes." The question before patient and physician alike was whether this cough was the sign of a disease. People who were sick seemed often as devoted to the theory of their ailments as any of their doctors—although it may have been a different theory. In addition, they often saw their doctor's theories about disease as a valuable basis for medical procedures and treatments. Outside of the doctor's earshot, of course, it may all have been very different. There has likely always been, as there is certainly today, considerable individual variation in the interpretation of suffering and illness. But the negotiation of Cabot's clinical practice took place—increasingly, perhaps, in the twentieth century—in the language of medicine and its related dialects. The experiences of illness that Cabot and his peers represented as naïve symptoms were commonly discussed and interpreted by patients in terms of disease. It is the persuasiveness of this shared model, and its application and its limitations in practice, that I consider here.[50]

The detailed records of Cabot's clinic have made me hesitate before seeing scientific medicine in the early twentieth century as something deployed unilaterally by physicians among passive and uncomprehending patients. Furthermore, the correspondence of these patients has led me to consider how the reception of scientific medicine shaped the persuasive influence of the physicians who advocated and deployed it. The power of technical medical practice seems neither to have been granted outright as a necessity of the doctor's work nor generated simply as a by-product of medicine's successful bid for professional autonomy. In Cabot's clinic, this authority developed in part through the interests of patients as they were expressed in a growing and densely organized twentieth-century market for medical services.

As patients moved in, through, and out of this clinic, on their own initiative and with the advice of others, they marked out channels through an expanding system of interrelated practices in and around Boston. This network manifested itself in different forms. It took shape professionally through the relationships among different physicians and practitioners. Cabot exchanged patients with colleagues through an intricate etiquette of referrals and consultations. Patients reached and left his office moving along links between doctors that revealed a traditional nineteenth-century collegial structure being overlaid by the demands of an ambitious twentieth-century profession. This network existed in an economic form as well, in the agreed-upon arrangements for calculating patients' bills on a sliding scale, taking into consideration their life circumstances. In setting their fees, referring and consulting physicians traded information about their patients' social positions. The same network manifested itself in the exchange of people and materials among these practitioners. The records of Cabot's clinic document the traffic of patients, blood samples, correspondence, family members, x-rays, pathological reports, bills, and payments among the offices, clinics, and laboratories of Boston and New England.

Identifying Patients: Methods and Limitations

In addition to paging through the thirty-seven volumes of Cabot's clinical patient records innumerable times, I also attempted to examine these records systematically through formal statistical sampling, in part to gain a representative sense of the patients and the practice. To acquaint myself with the collection generally, I first read a well-defined convenience sample of six hundred complete patient cases from the first patients in 1897 to the last in 1926, which provided a sense of the collection's organization and strengths. I then gained a representative picture of the practice by selecting a random sample of all the individual patient cases entered in these records during the main years of practice, 1900 through 1915. Cabot's careful methods for cataloging these records, once understood, assisted in creating the sample. Cabot entered his private clinical notes into the successive pages of blank, cloth-bound, folio volumes. For each new patient who came through his office door, he opened the year's active volume to a new page, entering basic information about the patient at the top of the page and reserving the bottom and reverse side for this and subsequent visits. On their return, patients who had been previously seen had successive information entered with their original record, until this page of the book was filled. When the record overflowed the two sides of the page, the continuation of the record in a different section was noted at the bottom of the page. At the end of each volume, Cabot, or more likely a secretary, created an index listing all the patients entered in the volume, allowing a returning patient's record to be recovered subsequently. Occasionally, the record-keeping methods slipped and a patient who had been seen previously triggered the creation of a new, unreferenced file in a later volume. But a card catalogue of all patients from these volumes attempted to identify patients systematically according to the volume and page number of their cases,

TABLE I.I   CASES LISTED IN THE INDEXES TO RICHARD CABOT'S
PATIENT RECORDS, 1900–1915

Volume	Year	Number of cases
1	1900	101
	1901	71
2	1901	16
	1902	138
	1903	97
3	1903	38
	1904	137
	1905	21
4	1905	259
5	1905	28
	1906	229
6	1906	204
	1907	21
7	1907	263
8	1907	174
	1908	148
9	1908	148
10	1908	175
11	1908	91
	1909	72
12	1909	178
13	1909	148
14	1909	146

(continued)

TABLE I.I *(continued)*

15	1910	148
16	1910	154
17	1910	145
18	1910	24
	1911	125
19	1911	142
20	1911	148
21	1911	58
	1912	90
22	1912	150
23	1912	142
	1913	7
24	1913	148
25	1913	150
26	1913	96
	1914	54
27	1914	150
28	1914	150
29	1914	55
	1915	95
30	1915	150
31	1915	74

Note: The organization of these records favored the use of the individual volume as the sampling frame, since volumes do not end consistently on the calendar year. Some volumes contain a greater number of individual cases, which influenced the choice of the method of sampling.

flagging these occasional oversights. Patients whose information appears in more than one volume without overt reference in the original record are relatively rare.[51]

Taking all patients listed in the indexes for the volumes from 1900 through 1915 created a total base population of 5,358 patient records from these busiest years of the practice, the great majority of them for different patients (table 1.1). From these records, I used random-number sampling to select 200 cases, corresponding to 200 different patients, a representative sample of the practice over this period. Given changes in the size of this practice over time, slight systematic variation in the number of cases in each volume, and volume breaks that tended to favor the year's end, I chose to use the individual volumes as the sampling frames and to sample from each volume randomly in proportion to the number of cases in that volume.[52]

For each selected patient in the random sample, I recorded the basic demographic data of the age, sex, occupation, and home city, which was available for almost all of these patients. In addition, I took down information on Cabot's diagnosis, the fee charged, and whether it was recorded as paid. I noted the presence of outside correspondence from colleagues or patients and reviewed it. Familiarity with Cabot's system of detailed record-keeping also allowed me to make a determination for each case as to whether the patient reached Cabot by one of three routes: either independently, on the referral of another doctor, or through a joint consultation during a meeting with another physician.

These representative statistical data captured the broad characteristics of the practice. Fifty-one percent of Cabot's patients were women, who had an average age of thirty-seven years, slightly younger than the average age of forty among the men. Patients were charged an average first fee of $13 by Cabot, with a median of $10 and a range from $0 to $100 in the random sample. Twenty-one percent of these patients gave their home city as Boston, with the majority of the remainder scattered across the Boston suburbs. Twenty-three patients in the random sample of 200 listed a home outside of Massachusetts, fourteen of them being from the nearby states of Rhode Island, Connecticut, Vermont, New Hampshire, and Maine. Roughly one-fifth of all the patients in the random sample came on the referral of another practitioner. Another one-fifth of these patients were seen first by Cabot in a formal consultation with their attending physician. The remaining three-fifths found their way to Cabot's office without the ostensible guidance of another doctor (table 1.2).

In understanding the nature of authority in this medical clinic, it was important to ask about the identity of the patients who patronized Cabot's practice. My sample gave reasonable information on gender, residence, and age, but where are social class, race, and ethnicity in this study of authority? These fundamental elements of social roles are necessarily inseparable from the perception and the expression of power. Yet in the archives of this clinic, these elements of personal identity are surprisingly obscure. Very few patients, for example, had their ethnicity or race identi-

TABLE I.2   DEMOGRAPHICS OF A RANDOM SAMPLE OF 200
PATIENTS IN RICHARD CABOT'S PATIENT RECORDS, 1900–1915

Number	200	
Average age (years)	38.7	
Female	104	
	(52%)	
Means of contact		
Consultation		44
		(22%)
Independent		118
		(22%)
Doctor referral		38
		(19%)
Average fee	$13.00	

fied by Cabot in their charts, partly because a uniform identity was otherwise assumed, but also partly because the differences were omitted. The majority of patients whose records note no ethnic or racial identity have Anglo-American surnames, but other patients who are identifiable through the details of their records as having different national and ethnic backgrounds have no identifying notes on this aspect of their identity. A patient who was noted incidentally to be celebrating Passover is not identified anywhere as Jewish. Other patients with ostensibly Jewish surnames similarly lack any ethnic identification in the chart; yet Cabot explicitly listed other patients in the records as Jewish in their identifying data at the top of the page.[53] Was this simply an oversight?

These records hint, in several ways, at an effort by Cabot to level the implications of differences in social identity, especially for race and occupational status. Disease mattered most to Cabot, and he gave information on diagnosis more consistently than almost any other piece of identifying information. Jewish identity, in fact, sometimes shows up only as a distinguishing feature of certain "Jewish types" of nervous disorder that Cabot diagnosed, discussed in detail in chapter 5. Differences in occupational status are similarly obscured in Cabot's notes. A man identified only as a piano maker in the chart, for example, turns out to have been the treasurer of a large company that made pianos.[54] Scant information about social roles was noted in the context of this office practice, outside of information significant to the obligations and roles as doctor and patient. Yet if this apparent effort to neutralize the significance of social identities seems artificial, it equally seems to have had an effect. These archives show evidence of an artificial space created inside the clinic for interactions between physician and patient, which minimized the significance of several obvious and common markers of social status.

The analysis here will, on occasion, read the significance of social identify back into these documents, especially in chapter 5, on nervous disorders, where the social identity of patients proved a powerful consideration. Yet the patients too conspired in their correspondence with Cabot to push aside considerations of social identity and role, barely hinting at their personal lives in their letters, and rarely and only indirectly using the evidence of their gender, social state, or ethnicity in their requests and descriptions. A patient whom Cabot identifies as a "former farmer" wrote to him to say simply "you were wrong," without further qualifications or apology, and then continued his correspondence months later, apparently with Cabot's encouragement. These correspondents acted somewhat analogously to the girls in the early twentieth century studied by the historian Joan Jacobs Brumberg. As the century progressed, Brumberg notes, girls as they began to menstruate spoke less and less as their mothers had about assuming the role and status of women and more as their daughters would about personal hygiene products.[55] In their archived letters to Cabot, patients speak scarcely at all about the connections out from their illnesses to their social lives and obligations. They seem intent, rather, on obtaining the proper medical attention and service. Further, they show a facility at times in reproducing the language of therapy and diagnosis of medical charts, which excluded much independent role for gender, ethnicity, or social position in interpreting medical problems or their care. It was a language that proved efficient for communicating a need for technical medical care, but one that lacked a terminology for expressing the meaning of these services to the social life of the individual.

Such impersonal articulation of medical authority may have had a special attraction for someone like Cabot, who saw himself as a reformer in an era of progressive reform. In a rapidly changing city like Boston, many of the previously reliable social links of neighborhood, kinship, and personal association were being weakened and overlaid. The urban reformers of this era sought new sources of organizational power in professional expertise and technically derived knowledge.[56] Medical authority might stake a parallel claim to control over matters of disease, which seemed independent of social identity and status. Cabot and his colleagues were heirs to a way of doing medicine that accepted the human body as a physical object but had leveraged the influence of individual physicians' practices through intimate knowledge about the lives and circumstances of the patients. Growing confidence in the categories of biological disease enabled Cabot and his peers to back gently away from an expertise grounded in knowledge of personal life, with its implications of a social authority resting in part on such characteristics as economic status, race, or gender. Physicians sought to simplify and narrow their claim to privileged knowledge of disease through the deployment of a technology that supported such claims.

Richard Cabot is still cited today in a number of present-day ethical debates as a medical muckraker who took on controversial topics that his colleagues typically hoped to avoid. Even as he documented the evidence of his patients' reactions to his practice, he debated publicly how physicians should presume to influence their patients and about how the mechanisms of such influence might impinge on the legitimate authority of his profession. As a national figure in academic medicine, he wrote provocatively on the misuse of placebo medications, the need frankly to diagnose the absence of disease in patients who wrongly thought themselves sick, and the obligation to disclose the diagnosis of a life-threatening disease when found. Cabot pursued and carefully recorded evidence of certain profound challenges that he met in deploying a technical, disease-oriented medicine in private practice. Although rarely representative of his colleagues in his personal views, Cabot left behind archives that provide unrivaled insight into the dilemmas of an early technical medical approach to disease. These archives illuminate the tension between the practices that a biomedical model of disease could readily accommodate and the practices that physicians and patients would readily endorse. It is a tension that remains largely unresolved.

CHAPTER 2

ORGANIZING A PRIVATE OFFICE

Between Home and Hospital

On a late December afternoon not long ago, I found my way to the doorstep of 190 Marlborough Street in Boston. One hundred years earlier, this address had been the home and the private medical office of Richard Cabot. The building that I peered up at that afternoon through a scattering of snowflakes was a narrow brick house with a slate roof that cannot have looked very different in Cabot's time. Set flush into a line of similar nineteenth-century row houses, it opened almost directly onto the street. In a few steps, I could have knocked on the door and perhaps met the occupants. Instead, I turned and walked east down Marlborough Street through the neighborhoods where Cabot lived and worked for several decades after he first began practice in 1896. I was following in the footsteps of one of Cabot's patients, a woman who visited him in January 1902, perhaps on a similarly chill and dusky day.

Mrs. Moore, as I shall call this patient to preserve the privacy of her medical records, was about forty when she visited Cabot's office.[1] Her trip was not a long one, because she lived in Boston herself, and she returned to visit several other physicians besides Cabot that year. In one evening, I was able visit the addresses of all these physicians, whose offices had been in the same Back Bay neighborhood, in the blocks around Marlborough Street bounded by Boston Common, the Boston Public Library, and the Charles River. How did such a network of a private office practices grow and sustain itself in this neighborhood, and what kinds of collaboration and competition linked the individual practitioners? Mrs. Moore's path through the streets of Back Bay one hundred years earlier had traced some

The corner of Marlborough Street and Exeter Street, Boston, ca. 1880. Cabot's office at 190 Marlborough would have been just off of the left edge of the frame. Courtesy of the Boston Public Library, Print Department.

of the boundaries and associations of Cabot's medical community, outlining the local geography surrounding and supporting a single office.

Lined with marble stoops and stretches of brick sidewalk, Marlborough Street offered a quiet and elegant venue for my walk. Parts of it had not changed much since Mrs. Moore's day; gas streetlights came on as night approached, but with a look of feigned old-fashionedness that was surely an imitation of 1900 rather a remnant of it. Since its reclamation from the marshy banks of the Charles River in the 1860s, Back Bay had long been home to many of Boston's wealthier citizens. The blocks that I walked east from Cabot's office offered me a peacefully unbroken line of brick and cut-stone row houses. In 1902, Mrs. Moore had visited the office of an ophthalmologist here soon after she saw Cabot. The other direction west up Marlborough Street had taken Mrs. Moore on another day to the house at number 226, where she visited the surgeon and otolaryngologist Dr. Eugene Crockett. When I reached the next intersection at Clarendon Street, I turned left and walked one windy block toward the Charles River to reach Beacon Street and the house of another otolaryngologist, Dr. Joseph L. Goodale, whom Mrs. Moore also saw later that same year. Along the way, I passed one house marked with a bronze plaque listing professional offices inside. Discreetly visible from the sidewalk, the plaque announced the practices of a licensed social worker, several PhD-certified counselors, seven MDs, and a group calling itself Trauma Recovery Associates. One hundred years after Mrs. Moore's journeys through this neighborhood, it remained a place where you might go looking for therapy and advice. Yet much about medicine and medical practice had changed, even during Cabot's lifetime. What guided people like Mrs. Moore through these streets in 1902, and what did they find to hold their allegiance inside the parlors and offices of Dr. Cabot and his colleagues?

The success of Cabot's clinic rested on meetings and exchanges with patients, to whom the attractions of twentieth-century technical medical treatment may just have been becoming evident. They needed guidance at times in engaging physicians who offered such services. The records of Cabot's practice provide a remarkable window into the routine structures of medical office practice in the early twentieth century. It is possible to map the routes that led people like Mrs. Moore to the doorstep of 190 Marlborough Street and the routes that led them away and into the offices of other affiliated doctors and healers in the New England of the day. The development of this clinic suggests how a network of scientific and technical services captured the patronage of a segment of people who were out searching for medical assistance.

Cabot and his peers developed office practices that were a hybrid between established, nineteenth-century medical arrangements and the novel forms that would increasingly define medicine in the later twentieth century. These practices gave the familiar problems of disease a persuasive technical reality that drew people away from home and family, but that also provided reassurance for those anxious about being cared for outside the traditional context of domestic treatments and healing. Medical neighborhoods and office practices that were organized around technical services made the problems of being sick efficiently expressible outside of the

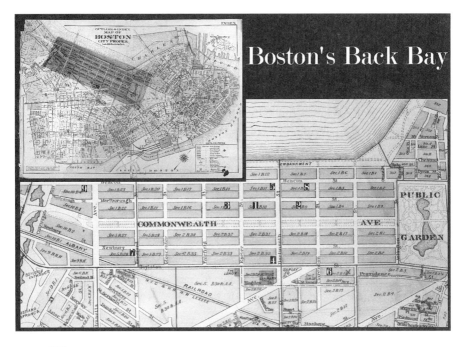

Boston's Back Bay

KEY:

1. 190 Marlborough St., office of Dr. Richard Cabot
2. 130 Marlborough St., office of Dr. Edward Reynolds
3. 226 Marlborough St., "Mrs. Moore's" first otolaryngologist
4. 700 Boylston St., Harvard Medical School (1883)
5. 259 Beacon St., Dr. W. J. Dodd's private radiological lab (1915)
6. 541 Boylston St., Drs. L. C. Rood, C. J. Alexander, M. F. Austin,
   B. A. Denig, and H. E. Rice
7. 51 Hereford St., Dr. F. L. Burnett's laboratory (1911)
8. 205 Beacon St., clinical laboratory (1914)
9. 499 Beacon St., office of Dr. Mary Dakin.

Boston's Back Bay neighborhood. Enlarged from Outline and Index Map in the *Atlas of the City of Boston—Boston Proper and Back Bay—from Actual Surveys and Official Plans by George W. and Walter S. Bromley* (Philadelphia: G. W. Bromley, 1908). Layout by Matthew Scanlon. Boston Public Library / Rare Books Department. Courtesy of the Trustees. Unless otherwise specified, the addresses are ca. 1906.

home, in part through a set of concrete, material exchanges. Practices like x-ray photography, surgical excision, pharmacotherapy, and therapeutic uroscopy translated the basic needs of sickness into a set of easily articulable requests that patients and physicians alike could act on, although they remained within the bounds of a comfortable model of personal medical service to families in domestic settings. Relying on the assistance of his colleagues, Cabot channeled his patients into the offices of a growing range of specialized physicians and practitioners providing these and other such services. Still, Cabot practiced, like many of his neighbors and peers, out of his own home, and visited people in their homes on occasion to provide care. Patients paid directly for their care in this practice in accordance with a sliding scale of charges that promised to take account of their personal circumstances, so that fees were matched to the individual ability to pay. Office practice fitted into a growing market for urban specialized medicine; yet it was also a market that remained personalized and was regulated through the relationships established between individual doctors and patients.

Disease was certainly the focal point of Cabot's technical management. But the management of disease had to match existing expectations about medical care. Domestic care, individual circumstances, and personal service were features of a model that focused on disease but recognized that patients had troubles that extended beyond the narrow confines of technical medicine. Cabot thus sometimes directed patients into a parallel network of spiritual advisors, mental health therapists, counselors, and charitable societies for additional support. While his overriding consideration was the identification and management of disease, his response to any individual patient was flexible. It was a style of practice that succeeded in drawing in, supporting, and motivating the many people who came seeking the resources of scientific practice in Back Bay medical offices.[2]

Medical Tools and Specialized Practice

By his own later account, Cabot developed his private office practice initially through the promotion of his specialized skills in the examination of blood samples for colleagues around Boston. But laboratory skills with the microscope and test tube offered only a shaky foothold for a young practitioner. These technical practices were rapidly evolving and shifting in their significance for routine office practice. The course of Cabot's own early career suggested a changing role for the laboratory in private practice. After graduation from medical school and a year's stint as a house officer on the wards of the Massachusetts General Hospital, Cabot began researching the use of white blood cell counts in the diagnosis and prognosis of conditions like pneumonia and typhoid fever. He obtained a year of support in 1894 as the Dalton Scholar at the hospital and pursued more advanced study of the microscopic and serological properties of blood in various disease states. His success was rapid and noticeable. By 1904, his *Guide to the Clinical Examination of the Blood* (1896) had gone through five editions, becoming a fundamental textbook in American clinical hematology and drawing the attention of colleagues both locally and around the nation. Cabot later noted that 1896 had also been the year "I hung my shingle out." He had begun by doing "most of the blood examinations for other doctors"; indeed, he later recalled that "the only way I got cases was through other physicians."[3] Specialized laboratory skills thus provided a route into practice for this novice physician.

With his publications gradually appearing in medical journals and his appointment as physician to the Out-Patient Department at Massachusetts General Hospital in 1898, Cabot began to attract the attention of other doctors, some of whom sent patients to him. "I am desirous that you make a complete blood examination and kindly send your report to me," a letter referring one patient reads.[4] As his reputation grew, doctors also wrote praising Cabot for his research and explaining that it had been an incentive for seeking his assistance. Referring physicians also began to send their patients to Cabot's office not just for blood testing but with general questions about the nature of their disorders, leaving it to Cabot to decide about their evaluation and care.

Specialized blood testing was limited as a means to sustain a private clientele. Testing blood was not the same as providing medical care. A specialized technical practice could cement good relations with colleagues, securing a role in providing access to an interesting, new procedure, for example. But laboratory practice made one's access to patients dependent upon the courtesy of colleagues. Some physicians, like Dr. Frank Day, wanted only the results of Cabot's blood examinations and were perhaps reluctant to send him a private patient when a blood sample might suffice. Day wrote to Cabot in 1899 requesting, "[C]an you help me out any by giving me your inferences from the blood sent in by same mail (both on cover glasses and on paper)?"[5] Cabot did not provide this requested blood test, although he was willing to exchange technical information on the blood counts, on one occasion, for example, sending samples of the special stains that he used to differentiate white blood cells.

The referring letters from other physicians occasionally pointed to differences between Cabot's academic position and the more work-a-day world of conventional private practice. One Boston physician sent a man with a "glandular swelling" to Cabot's office, who arrived bearing a letter stipulating that "as he is a patient of mine I should like to know the results of the blood count . . . and am glad you have time to do it."[6] This colleague suggested that Cabot's skills were a service to him as much as to his patient, and that it was a service perhaps better left to someone free of the demands of a busy office full of patients.

The significance of special laboratory skills to medical careers had been in transition over the preceding decades in this same medical neighborhood. The arrangement under which Cabot came on the medical scene—examining blood samples for other physicians—was ceasing to be a viable enterprise by the early 1900s. A comparison with the private practice by another physician a generation earlier in this same neighborhood illustrates the changing role of laboratory practice in private medical careers. William Whitworth Gannett got his start in medicine in Boston almost two decades before Cabot in 1880 in a very similar way, providing specialized laboratory analysis to his local colleagues. Gannett's private office opened at 110 Boylston Street, about seven blocks from Marlborough Street on the Boston Common, where he initially shared quarters with an established senior physician, George Tarbell. Gannett was also a graduate of Harvard Medical School, and he obtained the same entry position in the Out-Patient Department at Massachusetts General Hospital that Cabot did. He began by offering to analyze specimens of urine and blood microscopically and chemically for a fee. [7] But Back Bay had hosted a smaller and more personally interconnected medical community in Gannett's day.

The exchange of fees, reports, and patients between Gannett and other doctors depended heavily on local social and neighborly connections of a sort that were becoming attenuated by the time of Cabot's entry into medicine. Account books and laboratory notebooks from Gannett's office show a tight network of collegial exchanges that fostered his start in practice.

One of his regular supporters was, for example, his office mate Dr. Tarbell. Professional relationships for Gannett were personal and close to home. Tarbell was both host and client, and when Gannett moved out from Tarbell's office, he took space in the office of Dr. Arthur Cabot, a relative of Richard Cabot's, who apparently took on a similar role as a local patron. Arthur Cabot might drop off a patient's specimen in Gannett's office in the morning with a note asking, "Dear Billy, will you examine this urine to see if there is any trouble in the kidneys?"—and in the evening look in for the results. Another Boston physician, Dr. Morse, left a specimen with a note asking Gannett to reciprocate by stopping in later to deliver the report on it. Dr. Chamberlain came over to Gannett's office to pick up his report but arrived before it was ready, as Gannett reminded himself in his account book, while Dr. Elliott took the precaution of leaving a note alerting Gannett that he intended to call on him the next afternoon for the results. Dr. Porter was fortunate enough to catch Gannett on the street on a day in May 1889 to ask him about a test that he had requested—a fact that Gannett also made note of in keeping his financial books up to date.[8]

These arrangements quickly established Gannett's reputation among a cadre of his colleagues, so that in his office he likely saw as many physicians as he did patients. Access to patients came largely through his medical patrons. His colleagues recognized his dependence on their business and were occasionally solicitous in their correspondence. Dr. F. G. Morrill sent Gannett his typical $5 fee for laboratory analysis of a patient's urine, along with a note promising to "see that you have another whack at this urine in the fall." Gannett's laboratory services offered a variety of attractions for the physicians who sent him their specimens. Perhaps they felt, as did Cabot's later clients, that they were too pressed for time to do this work; or perhaps they lacked Gannett's training and equipment. They clearly identified additional value in the ability to consult with this laboratory expert. In a short note in January 1888, Dr. M. Holbrook wrote to thank Gannett. "This patient for whom you made the analysis was sure she had Bright's disease," he explained, "and nothing would convince her to the contrary but this analysis by an expert." Along with his gratitude, Holbrook enclosed "a money order for $5."[9]

By the turn of the century, such laboratory testing services existed in larger cities around the country. The back pages of contemporary regional medical journals published in Chicago, Cleveland, St. Louis, and Charlotte,

North Carolina, for example, list professional announcements by physicians providing them. Arrangements differed slightly between different practitioners. Dr. M. O. Hoge of Richmond, Virginia, advised his clients to "send specimens PREPAID, by mail or express in a well-corked bottle,"[10] for example, while Dr. Jones of St. Louis noted that advice on shipping specimens and billing would be "cheerfully given" on request.[11]

The physicians in and around Boston who patronized Gannett's practice often sent the specimens and the payments themselves, collecting the fees separately from the patients, as their notes indicate. On other occasions, they had their patients deliver a specimen to Gannett and pay the fee directly to him. The one stable feature of their arrangements with Gannett was that the results of the analysis always went back to the initial private physician. Gannett seems not to have released his information directly to the patients. He did not attempt to circumvent his patrons, the other physicians. Since the referring doctor alone received the report, Gannett's medical opinions and advice reached the patients only indirectly through their principal doctor, leaving it to that physician to interpret their significance.[12]

These arrangements were more delicate because Gannett operated in his practice as a physician himself, aspiring to provide expert medical opinions based on his specialized skills with the microscope and test tube. He struggled to avoid the role of a minor technician, merely processing laboratory specimens for better-established colleagues. His patrons generally cooperated in allowing him to apply his medical skills. They often sent specimens to Gannett attaching a question of particular medical significance. The surgeon C. B. Porter, for example, wrote seeking "to know the condition of [his patient's] kidneys with regard to operation," and received the reply from Gannett: "no bar against operating." Gannett supplied to his patrons an opinion about the medical condition of a patient based on his analysis, rather than merely observations and measurements of the laboratory specimens. Appended to the report of the specific chemical and microscopic findings in the urine, he might report, for example, that "there is probably some general disturbance of nutrition, but there is no evidence of disease of urinary tract." At the end of each report that Gannett provided to his physician-patrons was a section headed "opinion," providing such interpretations. Gannett carefully acknowledged the limits of what he could discern simply from the analysis of a laboratory specimen, suggesting in a typical case, for example, "I should be unwilling to make that diagnosis [based] on the urine alone." He always offered an opinion, but it was frequently expressed with the caution due to a close colleague.

Gannett refused to cede clinical judgments about laboratory results entirely to his referring physicians, preserving a role for his own skilled judgments, and he eventually ascended out of a limited laboratory practice, even as the techniques of chemical urinalysis became more routine and standardized. Up until the final year of his laboratory practice, for example, he avoided reporting on the chemical constituents of samples in objective, quantitative terms, preferring to list whether the levels of urea or uric acid in a urine sample were "slightly diminished," "not increased," or "improved." When one physician wrote emphatically stating, "I am anxious to know *how much* the urea is diminished," Gannett replied only that it was "much diminished," although the techniques for quantitative estimates of urea were well established by that time and must have been well known to Gannett, who had studied medical chemistry and pathology in Europe for a year after medical school. Gannett tenaciously maintained his role as a medical diagnostician based on his laboratory expertise.[13]

Gannett's aspirations seem to have been to ascend from this technical laboratory practice in support of his colleagues into a general private practice of medicine, and he eventually did so. Coincident with first opening his private laboratory service, he took on a series of unpaid positions as pathologist at Boston City Hospital, Carney Hospital, and McLean Asylum, winning local recognition for his special technical skills, as Cabot himself would later at Massachusetts General. These positions of hospital pathologist in fact served in Gannett's day mainly as a stepping-stone to appointment to the regular medical staff. Gannett finally made this step up in 1891, when he became a staff physician at Massachusetts General, where Richard Cabot would later join him. Over the next year, he began to shift out of laboratory practice into what was to be a long, thriving private medical practice in Back Bay, establishing himself sufficiently to give up pathology and begin doing the laboratory analysis for his own patients—or perhaps sending out specimens for other workers to provide the results.[14]

Although Cabot's specialization began in a way similar to Gannett's, by 1900, both the use of laboratory services and relations among practicing physicians in these neighborhoods had changed substantially. Colleagues occasionally asked Cabot to examine only a blood specimen, without the patient, even seeking similar arrangements to those that Gannett had permitted, with a fee collected separately and forwarded to him. Cabot, however, usually declined. Other physicians in the area were providing these services too. In 1905, Dr. James Lewis sent out notices advertising laboratory services from his office on Beacon Street in Back Bay, announcing his ability to analyze specimens ranging from blood, pus, urine, and sputum, with fees clearly indicated. Lewis promised interested colleagues in an attached letter that his reports would "be rendered by telephone at the earliest possible moment."[15] Cabot's private patients occasionally went three blocks from his office to the residence of Dr. Frank L. Burnett at 51 Hereford Street, who provided serological testing to patients for a fee in 1911. Burnett seems to have been following Gannett and Cabot in providing independent laboratory analysis out of his private office. But separate commercial medical laboratories were already to be found in Boston within the next decade. Burnett was actually among a last few in Boston to sustain this model publicly, and he too soon abandoned it. It became increasingly unusual for physicians in these neighborhoods to identify their individual private practices as sites for laboratory services. By 1914, Burnett himself had opened a separate commercial medical laboratory, called simply "The Clinical Laboratory," at 205 Beacon Street, while maintaining a general medical practice at a different address under his name. This was among the first of several independent commercial laboratories that appeared in Back Bay under such anonymous institutional names. Individual private practitioners like Burnett and Cabot simultaneously dropped the habit of identifying their private medical offices as sites for the analysis of medical specimens for colleagues.[16] With the increasingly common use and gradual standardization of the technique and instrumentation of these laboratory analyses, Cabot and his colleagues shifted routine testing to anonymous commercial laboratories.

While commercial laboratories provided new and difficult analyses, such as the Wassermann test for syphilis, individual physicians like Cabot also kept available in the office the equipment for simpler laboratory procedures. Cabot made use of a number of special tools and procedures in his office, complementing the resources available from nearby colleagues and laboratories. His key piece of equipment was a microscope with an array of special staining apparatus for identifying bacteria and the various types of white blood cells. Perhaps because of his special expertise in blood diseases, Cabot frequently left notes tallying the counts of the various lineages of white blood cells in the margins of his office clinical records, confirming that he had performed the counts in the office, apparently sometimes while the patients waited. Cabot also kept basic apparatus for chemical urinalysis, of the kind that Gannett had used in his laboratory practices in the 1880s, including a device for measuring specific gravity, chemical reagents, and glassware. For blood analysis, he also used hemacytometers and several different types of colorimetric devices for estimating degrees of anemia, called hemoglobinometers. Among the basic tools that Cabot routinely used, many served diagnostic purposes. The stethoscope, the reflex hammer, and the blood pressure cuff were the most common. He must also have had a gynecologic speculum and a table for female pelvic examination, because he sometimes noted findings from examining the cervix. He also kept equipment for minor therapeutic procedures, such as trochars for tapping fluids and minor surgical equipment for incising abscesses and changing out surgical drainage wicks, judging from the notes of procedures performed in his office.[17]

The range of available equipment, procedures, and tests grew rapidly during the first decades of Cabot's practice, giving him access to a growing range of radiological procedures and serology. X-ray evaluations were available by 1911 from Dr. Walter J. Dodd and his associates, who operated a private radiological laboratory at 259 Beacon Street, just around the corner from Cabot's office. Also housed in this large nineteenth-century apartment building were the offices of six other physicians, including two who offered specialized obstetrical and gynecological services. A surgical colleague, Edward Reynolds, performed cystoscopies and other minor urological procedures, often operating in his office just a block down Marlborough Street from Cabot. One of Cabot's patients, whose remarkable story figures in chapter 4, followed Dr. Reynolds from this office to his operating suites in a pair of connected row houses a mile to the west on Massachusetts Avenue, the site of the recently opened New England Deaconess Hospital. The Massachusetts General Hospital also provided access to a growing array of special treatments and tests, both for patients in need of procedures not available in the local offices and for those without the means to afford them, since the hospital provided free care for people judged unable to pay.

A Network of Domestic Practices

A medical office like Cabot's lay midway between the domestic world of nineteenth-century private practice and the twentieth century's fast-growing institutional medical realm.[18] When patients came to 190 Marlborough Street, they were entering not only a complex network of specialized practices but Cabot's private home. The surrounding Back Bay neighborhood was a mix of residential and professional buildings, including the offices, labs, and homes of Cabot's medical peers and other elite Bostonians. The geography of these practices reflected a professional medical world in the midst of substantial change. Physicians were reorganizing the spaces of medical practice, and the medical office was emerging as a crucial institution.

Harvard Medical School seems to have been the origin of this vigorous medical district. In 1883, the school moved from its location on the east side of the city, near Massachusetts General Hospital, across town to 700 Boylston Street, in the heart of the Back Bay neighborhood, one block from Copley Square, where the city broke ground for a new Boston Public Library a few years later. This exclusive neighborhood in the city's center had already been developing for almost two decades, and the large landfill project that had created it was completed in 1886. Expensive new row houses on these streets were the only new houses being built inside the city from the 1860s through the 1890s that appealed to wealthy buyers, and so they quickly drew residents from among Boston's elite merchants and business owners. It must have seemed an attractive location to live and practice to the well-connected physicians who located there toward the close of the nineteenth century. Among these, as we have seen, was William Gannett, whose laboratory practice in the 1880s relied on the patronage of physicians connected to the medical school. When the medical school moved again in 1906, to a newly built campus on Longwood Avenue at the western edge of the city, it left behind a thriving community of practitioners, who maintained their offices in the area.[19]

The arrangement of medical practices in this neighborhood shows the influence of the school, and something of a local hierarchy based on medical-school seniority, even after the school's move out west to the Longwood campus. In 1906, the two blocks between Cabot's office and the office of the surgeon Edward Reynolds at 130 Marlborough Street included four other row houses that held the practices of five other physicians, all of them graduates of Harvard Medical School and two of them with active appointments on the faculty there. All five of these physicians were, like Reynolds, senior members of the community, who had graduated from Harvard Medical School between 1858 and 1879. Among these five, three occupied separate row houses with their private offices-and likely also their residences-while the two physicians who shared an address were practicing in the related special surgical fields of ophthalmology and otolaryngology. Their younger colleagues on the next block managed less grand accommodations, however. At 129 Marlborough Street, five physicians shared a single three-story building for their offices, either lodging in smaller apartments in the same building, or commuting into the neighborhood for their practices. All five were also graduates of Harvard Medical School, of more recent matriculation, between 1880 and 1901.[20]

If Harvard Medical School was the stimulus for this medical neighborhood, it did not limit its constituency. These blocks of fine real estate held other obvious attractions. Doctors and practitioners from a range of backgrounds flocked to Back Bay addresses. The building at 541 Boylston Street near Copley Square, for example, held the offices of five physicians listed in the Boston Medical Directory for 1906. One of them, Dr. Luther C. Rood, like his colleagues a few blocks over, was also a graduate in 1899 from Harvard Medical School. Other physicians in the same building, however, had trained in several other very different institutions. A group of women physicians also in the building included Clara J. Alexander, Mabel F. Austin, and Blanche A. Denig, all affiliated with the New England Hospital for Women and Children in the western suburb of Roxbury. These physicians were graduates respectively of Women's Medical College of Pennsylvania, Johns Hopkins University School of Medicine, and Northwestern University Women's Medical School. Harvard Medical School did not confer medical degrees on women until decades later.[21] In the same building at 541 Boylston Street was also the office of Harry E. Rice, who listed himself as a graduate of the New York Homeopathic Medical College and a member of the Boston Homeopathic Medical Society. Other office buildings in the neighborhood also mixed similarly diverse groups of physicians. During the same period, 220 Clarendon Street, at the northeast corner of Copley Square, housed eight physicians, two graduates of Harvard Medical School and six others associated through membership in the

Massachusetts Homeopathic Medical Society, with training at homeopathic medical schools in New York, Philadelphia, and Boston.[22]

The medical profession had begun to contract and to raise barriers in
education and licensing to limit access, but it still kept the broad contours
of a less exclusive nineteenth-century constituency. To a medical patriarch
like the Boston physician Oliver Wendell Holmes in the mid nineteenth
century, homeopathy was a competing, and illegitimate, form of medical
practice, one among a number of "kindred delusions." Homeopathy
nonetheless continued to hold the well-established status of a viable, well-
recognized form of medical practice in late nineteenth-century Boston.
Harry Rice and the other homeopathic practitioners with offices at 541
Boylston all held MD degrees from their medical schools and maintained
listings in the 1906 register of "legally qualified physicians" published by
the American Medical Association. Such routes into approved medical
practice were rapidly closing down in Cabot's day, however. The fates of
the alma maters of the physicians at 541 Boylston Street illustrates the general trend. Dr. Rice's homeopathic New York Medical College did not sur-
vive the reforms of the early twentieth century. Dr. Denig's Northwestern
Women's Medical School closed in 1902, and Dr. Alexander's Women's
Medical College of Pennsylvania only narrowly escaped the fate that befell
all other traditionally women's medical schools in the United States during
this period. Boston University College of Medicine, the alma mater of one
of the practitioners at 220 Clarendon Street, successfully shed its identity
as a homeopathic college in 1918 and simultaneously limited the admission of women students, illustrating the reinforcement between these parallel limiting trends.[23]

Markers for conventional medical practice existed in the early twentieth century in diplomas, society membership, and licensing qualifications.
Reforms of the period would gradually tighten and more tightly enforce
such regulatory mechanisms. But the constraints on practitioners seem
not to have limited the movement of patients much at this time. Records
from Cabot's practice show that his patients sometimes traveled quite
widely among an eclectic array of practitioners. One woman who wrote in
1902 had recently returned home from a visit to a doctor at a special sanitarium in Buffalo, where she had been treated by electrolysis. She reported
on care that she had received variously from four different doctors in Buffalo and in Massachusetts, as well as an unnamed "herb Dr.," whose locale

she did not specify. The patients frequenting Cabot's office, considered as a whole, seem to have been a generally adventurous group, open to many different modes of therapy. They described their experiences pursuing care variously from "an osteopathic treatment every week" to homeopathy, herbalism, and Christian Science. Cabot and his specialized medical peers occupied only one corner of the health-care marketplace.[24]

Some patients apparently stuck to visiting only physicians with formal medical certification, but they often saw many different specialists. Cabot's office sometimes served as an intermediate stage in a long journey through medical offices in and around Boston. A given patient's record in the files might open with a recitation of what and whom this person had already tried for their troubles. One middle-aged man suffering from back pain arrived at Cabot's office in 1909 bearing such a story. He had already sought treatment from a professor of therapeutics at Harvard Medical School, who had prescribed "aspirin and potash." After four months of this treatment, he went to two orthopedic surgeons, the attending physicians at Carney Hospital and Massachusetts General Hospital respectively, who shared an office at 372 Marlborough Street with a third, more senior surgical colleague. These surgeons "strapped him & corset & rest," but this course of treatment turned out to be "no good." The same orthopedists next "etherized [him] & stretched sciatic nerve & broke adhesions," which helped briefly, according to Cabot's notes on the case. This man then got a steel brace to help him walk and went back to the same two surgeons, who called in their senior office mate for another opinion. But an unfortunate setback derailed their plans, apparently. While in their office, this man was bending over when suddenly "sciatic returned & is now awful," Cabot noted. The surgeons—at this point in Cabot's notes, their identities become blurred—treated him with aromatic ammonia, "wh[ich] gave great relief at first." But after five months more, they "said go to Ellis," referring him on to a general physician, who helpfully provided morphine for his pain. Then, as the record noted, "McBurney of New York saw him at Stockbridge," in western Massachusetts. Finally, he consulted Cabot, who diagnosed his condition as sciatica and advised him to return to the original orthopedic surgeons for further treatment. The record ends there.[25]

As this account suggests, the medical office was an important institutional base for specialists. Urban hospitals were increasingly important in the careers of Boston practitioners, and surgeons were easily identified by their hospital affiliations. But the medical office was a way station between home and hospital. The man with the back pain visited many offices, but he does not seem to have been hospitalized. Medical offices had long been a convenient extension of the physician's primary residence, a place to proclaim one's professional status and gather patients. In Cabot's day, however, the medical office was increasingly not just an extension of the physician's residence but an independent workplace in its own right. The daily routines of some of Cabot's colleagues illustrate this. In a 1904 professional guide, an enterprising ear, nose, and throat man announced regular office hours from 9 A.M. to noon daily in Beverly, Massachusetts, and then again from 1:30 to 4:30 P.M. at an office a few miles away in Salem, bespeaking not only his mobility but his reliance on his medical offices. The surgeon Frank Balch lived on Clarendon Street, very close to Cabot in Back Bay, but had his surgical offices farther to the west at the intersection of Massachusetts and Commonwealth Avenues. Conversely, John Brainerd, a specialist in diseases of the ear, nose, and throat, who lived out in the western Boston suburb of Brookline, listed his office as being at the Hotel Copley, near the Boston Public Library.[26]

The Development of Office Practice

In the mid-nineteenth century, the medical office had been a place where patients, family, and physicians met on the way to the patient's home. Home visits had been the cornerstone of American medical practice before the turn of the century. Although home visits retained a powerful symbolic value for the medical profession through most of the twentieth century, their practical significance seems to have been in the early stages of a gradual decline by the first decades of the century, as the institutional basis of practice shifted. In the mid twentieth century, the medical office became a place where physicians and patients met on the way to and from the hospital. Dr. Daniel W. Cathell, a popular late nineteenth-century advisor of American physicians in matters of "professional tact and business sagacity," recommended that the medical office should serve as a place to gather patients who would become the basis for a stable practice of home visits. Cathell's book *The Physician Himself and Things That Concern His Reputation and Success* had remarkable popularity from its first publication in 1881 into the reprints of the 1920s, offering reams of pragmatic advice to the aspiring private practitioner. Cathell counseled that regular office hours were useful to the young physician in order to pick up the "overflow, emergencies, cases of accidents, calls from those who are strangers in the city and other anxious seekers," which other established physicians would miss in their daily rounds to their patients' homes. A contemporary of Cathell's, the surgeon Arpad Gerster, attributed his start in private practice in the 1870s in New York City to just such an emergency visit from a patient who had found other familiar physicians away on house calls. Cathell assured his young medical readers that office hours could become more limited as their practices picked up and they came to rely less on the "transient office-patient." He reserved the bulk of his recommendations for the cultivation of the role of a family doctor through proper comportment on home visits.[27]

As a transitional space linking domestic and institutional worlds, the medical office in the early 1900s presented the patient with a mix of impressions. When patients came to see Cabot in his office on Marlborough Street, they entered the first floor of his home. In describing their contacts with him, they used terms occasionally that were appropriate to a personal visit, like the man who wrote in 1911 saying, "I called upon you a little more than a year ago," or another man who recalled having "paid a visit to you on April 2."[28] Other physicians also tended to refer to the medical office as identical with the place of residence, like the doctor who wrote to say that he had referred a patient to Cabot, "but unhappily, you were not home at the time."[29] Other people, however, seemed uncertain about the connection between home and office. One woman wrote to Cabot in 1906 at his home address asking, "[I]f I am to go to your office will you please state what days and hours you are in your office and give me your address?"[30] Similar uncertainties troubled a man who wrote seeking to have Cabot examine his urine—much the way William Gannett had for patients a decade earlier. He found Cabot unresponsive and inquired: "I sent to your address a small bottle of my water some time ago and got no reply from it yet and since I have sent it I have been thinking perhaps you would rather have it sent to the hospital." The shifting use of the medical office left patients unsure whether a physician's principal workplace was his home or some other place, such as a hospital clinic.

Many of the physicians in Cabot's neighborhood maintained a combination of home and office in one space. This mixing of domestic and professional space had various implications. Dr. Mary D. Dakin, an 1890 graduate of Boston University School of Medicine, maintained her private gynecological practices in a combined residence and office at 499 Beacon Street that she shared with Dr. Edward A. Dakin, presumably her husband, who had come to Boston from a homeopathic medical school in Philadelphia just a few years earlier.[31] Cabot's own combination of home and office followed the pattern of many of his medical neighbors. By the accounts of contemporaries, early twentieth-century medical practices in Back Bay row houses typically included a large first-floor medical office with family residences on the second and third floors above. One common arrangement consisted of a first-floor waiting room in the front of the house that opened into a large central "consulting room," with smaller private spaces for examinations off of it.[32] Younger physicians with less spacious accommodations lived in apartments connected to "make shift" medical offices in their same building, keeping their expenses low. As their practices grew, they sometimes kept their offices but moved to separate residences. But a separate residence was, as one physician who practiced out of the busy professional building at 259 Beacon Street later recalled, "a real innovation in those days."[33] The six homeopathic physicians previously mentioned who practiced at 220 Clarendon seem to have caught on to this innovation early, since five of them listed residences separate from their medical offices by 1906.[34]

Private medical offices at the turn of the century often seem to have struck a balance between a neutral institutional appearance and styles more expressive of domesticity. The head of the clinic, these spaces suggest, might equally be the head of the home. Advice manuals offered detailed recommendations about rugs, lamps, paintings, curtains, and wallpaper in the waiting area, taking into account the possibility that physicians might choose to keep the office separate from their residential space.[35] Daniel Cathell's popular advice manual enjoined the young physician in avuncular tones about the proper balance of domestic and professional decor in the office, including "diplomas, certificates of society membership, potted or cut flowers or growing plants or vines, fine etchings, pictures of eminent professional friends or teachers [etc.]."[36] This list alternates between the homey and the professional, positioning medical office spaces carefully along the continuum between traditional domestic healing and an updated institutional milieu.

Guides like Cathell's typically assumed a male audience and pitched their advice about the domestic nature of practice in a way that maintained a complicated relationship to male social authority in medicine. There was, for example, a common overlap in this guidance literature between the representation of the male physician in the office and that of the husband or father in the home. In his 1905 book *How to Succeed in the Practice of Medicine,* Joseph McDowell Mathews, the former head of the American Medical Association, advised the aspiring male physician to have his wife decorate his medical office—and then proceeded to offer advice on how to choose an appropriate wife for a medical career. The domesticity of the office should ideally reflect masculine oversight, rather than the male doctor's involvement. Cathell's manual cautioned against creating too much of a masculine look to an office space, giving it the appearance of a smoking room or hunting lodge.[37] A ready conflation of masculine forms of domestic authority with the physician's role extended widely through the culture of the day. An article by Cabot that was published in the popular *American Magazine* in 1916 included a half-page photo of him seated with his arm over his wife's shoulder, with two young girls playing quietly at their feet, seeming to show the doctor as the patriarch in his little family. This scene was openly contrived, since Cabot and his wife had no children. Someone at the magazine perhaps believed that this "distinguished Boston physician," as the caption read, should appear before his readers at the head of a traditionally pictured family.[38] The advice literature of the day suggested that physicians might adopt a similar kind of artifice in arranging the office, seeking to represent at once both the domestic and professional manifestations of a male doctor's authority.

Consultation and Referral

Cabot was meticulous in recording information about the doctors who recommended patients to him, permitting the determination in most cases of the nature of his relationship to these colleagues. As we have seen, a significant portion of Cabot's practice reached him through formal arrangements with other physicians. These arrangements reveal the foundations of a tradition of medical consultation overlaid with a newer system of medical referrals that fitted patients into expanding array of specialized practices. Consultation was a well-established nineteenth-century protocol for

Richard and Ella Cabot posed with their two nieces, in a photograph dated 1911.
A version of this image accompanied an article written by Cabot for the *American Magazine*, 1916, 81 (4). Courtesy of the Harvard University Archives.

sharing patients between different physicians that had been refined over decades of use. But Cabot and his colleagues also relied on a newer system of referrals to knit together the expanding, more differentiated American medical profession that was emerging in cities like Boston.

By sharing patients, physicians emphasized the collective nature of their professional enterprise and reinforced their joint authority. Referral and consultation, as mechanisms for sharing patients, expressed different forms of professional authority. Under the protocols of consultation, the attending physician invited the consulting physician to meet with him in the patient's home to confer about a course of care. This older, nineteenth-century practice emphasized a joint personal connection to the patient, cautiously negotiated between different physicians in the presence of the patient and family. Consultation modeled the service of the physician on the traditional analogy of domestic care. The doctor came to the home to visit the ailing patient, and when two physicians came, they arrived together to pool their expertise. Consultation fitted neatly into a world where physicians worked from offices in their homes and largely inside the homes of their patients. Nineteenth-century advice on consultative etiquette stressed the need for the appearance of accord and the importance of choosing carefully before admitting other practitioners to the legitimacy of consultation.[39]

Referrals instead tended to distinguish among physicians according to what they did, rather than whom they visited, or whose practices they associated with. In a referral, one physician sent the patient to another physician, often with an accompanying question about a special procedure or a problem to be considered. Patients coming on referral to Cabot's office sometimes carried, or were preceded by, letters from their physicians outlining the question. Referrals usually implied functionally distinct services, and the movement of patients outside the home between different medical institutions like private offices, laboratories, or hospitals. A system of referrals took shape partly through the definition and provision of specialized technical needs. The process of referral required practitioners and patients alike to resolve illness into discrete problems, each amenable to specialized intervention. Confirmation of the diagnosis might require referral for an x-ray to a physician providing radiological services. Treatment might require referral to a general surgeon or to a urologist or a gynecologist. The purpose of the referral also depended in part on the qualifications and skills of the referring physician. A specialized colleague might refer a patient to Cabot with narrowly phrased questions about blood disease, while another doctor might request only general medical care.

Referral and consultation taken together provided slightly less than half of Cabot's total practice. A large number of people joined his practice without any formal direction from another practitioner. About three-fifths of the individuals in a random sample of cases from the records of 1900 to 1915 made their way to 190 Marlborough Street independently, without the apparent guidance or advice of another doctor. They were sometimes brought there by word of mouth or happenstance or were attracted by Cabot's reputation. In the one-fifth of cases sent by referral, Cabot's records indicate the name, and often the address, of the practitioner who had sent the patient. The final one-fifth of the cases in the sample consisted of people whom Cabot saw in consultation with another physician, meeting together at the patient's home (see table 1.2).[40]

The physicians in Cabot's acquaintance displayed a ready familiarity with the arrangements for consultation and were generally attentive to its protocol. Cabot carefully noted when a patient was seen at the home "*cum* [with]" another physician in consultation. His colleagues generally tried to arrange with him in advance to meet together at the patient's home, although timing might prove difficult. One colleague wrote later after a consultation to chide him that he had not arrived in sufficient time for a previously arranged consultation. This out-of-town physician was the attending physician for a patient who was dying at home. "I tried to get you on the 29th but you could not get here until the following morning," he wrote Cabot after the patient's death.[41] Consultation had obvious limitations in providing timely assistance, although it held other attractions for the physicians who engaged in it.

Patients too demonstrated a familiarity with the methods of consultation, although they had their own questions about its details. One man writing to Cabot in 1914 recalled how Cabot "had been called in consultation with Dr. Davidson and Dr. Sweet of this City. Dr. Davidson the attending physician." Indeed, Cabot recorded that he had traveled to the patient's home outside of Boston the previous year for this consultation, "C[um] Dr. Davidson." The patient who met with Drs. Cabot, Sweet, and Davidson reminded Cabot that at the time, "you all agreed with Dr. Davidson's diagnosis of the case." He was concerned, however, that a degree of collusion had affected their deliberations. "Dr. Davidson told me," he informed Cabot, "that he had written several closely written pages of type written matter describing my life & case for your benefit before you saw me." He wanted to know: "Is that the common practice among physicians in consultation?" He seemed to wonder if the joint consultation had been preempted by too close an association among his physicians, impeding their independent judgments on his case. Although patients sometimes used consultation in this specific sense of the term, they also used it in the general sense of medical advice, as with the man who wrote, "[A]fter consulting you I returned home and have been doing as you advised."[42] This man clearly intended for consultation in this instance to refer to his visit to Cabot's office, as though he were the attending physician seeking an opinion and Cabot his consultant.

As measures of collegiality, the standards for consultative conduct were scrutinized and debated among American physicians in the latter part of the nineteenth century. Physicians in consultation were representatives of general professional conduct, meeting to confer together before the audience of their patients. The American Medical Association's 1892 Code of Ethics included an extensive entry, under Article Four, on "The Duties of Physicians in Regard to Consultation," specifying in ten numbered sections the details of proper behavior. Although the code was clearly intended as a guide rather than a description, the fact that the code was reviewed, debated, and modified repeatedly highlights its importance as an agreement about ideals for practice. Through its various iterations and revisions, the code spelled out what to do, for example, if the consulting physician arrived at the patient's home before the attending physician (wait); and how to respond if a patient requested a consultant who was not "considered a regular practitioner" (demur). Whatever the actual influence of the code, physicians clearly attended carefully to the management of consultation as a reflection of their professional duties and connections. One nineteenth-century surgeon recalled that among his colleagues in New York City, "the ceremony of consultative procedure was strictly maintained."[43]

During the period in the late nineteenth century when consultation still reigned as a defining element of medical conduct, debates over the proper etiquette for consultation echoed common concerns about the challenges facing physicians in private practice. The problem of competition among physicians was, for example, foremost in consideration. The two physicians who shared responsibilities for a patient in consultation ideally met in formal, face-to-face encounters. Consultation enforced some restraint among physicians, at least in theory, in vying for the patronage of a patient to the detriment of professional reputation. Anything done by one physician would be known to the other, and using consultation as a means of replacing the attending physician was forbidden. The American Medical Association's complete discussion of consultative protocol addressed in detail misdeeds like wooing away another practitioner's patients. The other professional controversy over consultation turned on the question of which practitioners could properly receive the benefit of collegial association. Again the Code aimed to restrain competition, in this case, by excluding practitioners identifiable as outsiders from professional association.[44]

During Cabot's decades in practice, the value attached to proper consultation gradually began to wane, probably tracking the general slow decline in the significance of house calls in private medical practice. The declining significance of consultation was evident to the physicians who lived through the early twentieth century. Consultative visits in the patient's home were still part of private practice in the 1930s but were a rapidly disappearing phenomenon. Dr. Dunbarton Shields, an internist in practice in Concord, New Hampshire, for the middle third of the twentieth century, recalled the change in his practice. He described how in the 1920s and 1930s, he had traveled to consult with colleagues in their patients' homes about heart problems, in the manner of his nineteenth-century predecessors. After World War II, however, his specialty practice shifted into professional offices adjacent to the central hospital in Concord. He and his partners now received their patients in the office clinic, sometimes on referral from the same outside physicians whom they had previously met with in consultation. Shield's son grew up to become a doctor too, and the father lamented that his son never once had the experience of making a professional house call.[45] The shift from consultation to referral figured in the experience of many physicians in widely different circumstances in the early twentieth century. Dr. Lewis Moorman, a practitioner in a prairie town, remembered a first difficult childbirth in 1902 that had required him to call in an older physician from another town, who arrived just in time to provide crucial assistance. Moorman recalled that by the 1920s, his medical office had been crowded with the patients referred to him by other practitioners, conveying the sense that cultivation of his office practice had been the key to his success, in line with Cabot's account of the growth of his Boston practice.[46]

As consultation diminished in significance, the American Medical Association's intermittently revised Principles of Medical Ethics reflected the emergence of referral as an important element of professional behavior. Focused in its earlier incarnations on the etiquette of consultation, this code had been silent on questions about referral in the versions of 1892 and earlier. The revision of 1903, however, included in the section on consultation a new set of stipulations for those unusual cases in which physicians who were sharing the care of a patient were not able to be present for consultation together. The code thus acknowledged that the traditional conference at the patient's home between attending and consulting physician might not always occur. The 1912 version went further, expanding the section on consultation to address the instance "when a patient is sent to [a physician] specially skilled in the care of the condition," that is to say, in the case of a referral.[47] Advice on the ideals of conduct had to keep pace with the changes occurring in daily practice.

Referral and consultation manifested different forms of medical authority for physicians and so presented different sorts of troubles in their management. Consultation placed professional authority on display under the guise of collegiality. Physicians gained reputation collectively in consultation by their ability to collaborate, to reach agreements, or to cede gracefully to one another. Essayists in medical journals cited the importance of maintaining the outward appearance of mutual respect and accord among consulting physicians, while acknowledging that the reality might not always be so pretty.[48] Both referrals and consultations supported the individual physician with the collective authority of the profession. In consultation, this professional authority was manifested as a collegial, interpersonal, and deliberative practice.

Referrals raised very different regulatory concerns. Guidance on proper conduct for referrals was largely subsumed into a larger, and often quite heated, controversy in the early twentieth century over fee-splitting, a practice in which a specialist, usually a surgeon, shared a portion of the fee paid by a patient with the referring physician. It was a form of collusion in which the referring physician was compensated—and so gained an incentive to refer—from the specialist's higher fees. In the form most widely denounced by physicians, fee-splitting took place without the patient's knowledge and often with some prearrangement between the specialist and the referring physician. Although physicians were generally happy to denounce the practice in this crude form, they did occasionally defend alternate versions, especially if they involved some kind of disclosure to the patient. Professional ethics generally forbade the practice outright.[49]

The differences between the problems of fee-splitting and the problems of consultation reflected the changing nature of professional relationships in the early twentieth century. The ethical regulation of consultation was concerned with creating formal cooperation and accord among physicians, whereas the regulation of referrals was concerned with preventing conspiracies, as, for example, in private collusion over the sharing of fees. A level of cooperation among physicians was already inherent in the practice of referring patients.[50] The increasing integration and consolidation of the medical profession in the early twentieth century shifted the nature of the challenges to its regulation. The complaints of Cabot's patients also occasionally shed an intriguing light on the complementary weaknesses in these two methods of sharing patients among physicians. When consultation succeeded, it brought physicians into close association. So the patient who was concerned about the written notes shared with Cabot by his attending physician, Dr. Davidson, suspected that his doctors were colluding. Patients encountered a very different problem when they traveled between referring physicians. Too little communication occurred, because referring physicians assumed that the physicians taking their referrals would assess the problems themselves. One woman, who was sent to Cabot's office by her main physician, Dr. Prescott, wrote to complain to Cabot that she was disappointed, saying, "I fancied that . . . Dr. Prescott had told you all the things connected with my case which I myself answered so inadequately yesterday." She worried that she was bearing too much responsibility in carrying the information between her two physicians. Her concern seems to find confirmation in other correspondence, such as a letter from a physician who wrote to Cabot to let him know tersely that his patient "will tell you about her present symptoms at greater length" when she reached his office on referral. Physicians seemed at times to value the independence of opinion allowed by referrals, writing to Cabot, for example, to "please give [the referred patient] advice as freely as you would if he were not my patient."[51]

The collective power manifest in referrals lay in the ability of each physician to do his or her special job well. The referring physician had no control over the actions of the physician who received the referral. Responsibility for choices about care passed to the next physician. So when a patient was referred to another doctor for a specific technical service, he or she did not necessarily get the service. Seeing a man with a draining opening in the skin under his arm, Cabot sent him to the Boston surgeon, Dr. Daniel F. Jones, who kept a private office around the corner at 267 Beacon Street.[52] Cabot apparently hoped that a surgical operation to excise the source of the drainage would remedy the problem. Jones wrote back after the visit thanking him "for referring him to me" but advising "a few months of letting [the opening] alone." Cabot, who had wanted the surgery performed, wrote on the chart: "Jones did nothing."[53]

Referrals assumed tacit cooperation among physicians. Cabot generally required only minimal arrangements by the referring physicians to receive patients. Dr. Walter Sawyer a general practitioner from Fitchburg, Massachusetts, sent a polite handwritten note to Cabot on his letterhead in May 1910: "I have a patient whom I would like to refer to you. . . . He will come to Boston to see you at your office at whatever time it is convenient for you."[54] Sawyer had apparently discussed the arrangements with his patient at least briefly before writing. "Would Thursday or any day in the latter part of the week be satisfactory?" he asked Cabot.[55] The arrangements and preparations could on occasion be more elaborate, however. The prominent Boston neurologist James Jackson Putnam saw a man with seizures on referral from another doctor in 1913 and sent over a carbon copy of the original referral letter for Cabot to have on hand when he saw the patient.[56] Some patients came bearing special personal appeals, as with the woman who brought a letter stating, "This will introduce to you my friend . . . [who] has been sacrificing herself in the fight for social justice in progress on the pacific coast. She is entitled to the best consideration that we can show her."[57] In this case, the referral was based on a personal, or really political, connection with Cabot, unrelated to technical medical services. Other referrals were accompanied with less ceremony or import. For example, Dr. Sarah Bond, a graduate of Boston University, who had an office on Boylston Street in Back Bay, not too far from Cabot's, in the 1910s, wrote simply to say that she was "desirous that [my patient] may have the benefit of your opinion." No special arrangements were requested, beyond that "she will go to your office tomorrow—Saturday."[58] The understanding seems to have been that the patient's willingness to present herself at the office when Cabot was in and perhaps wait until he was free would be sufficient to get a visit.

As specialized referrals among doctors in Cabot's circles overlaid existing arrangements about consultation, they occasionally created interesting hybrids. Some patients and their doctors wanted to meet Cabot in his office together, combining referral and consultation. One Boston doctor wrote to Cabot in 1906 "to ask if you can make an appointment to have [my patient] come to your office, at which time I would come with him as also would his sister." The arrangement was perhaps a bit unusual, and so the writer continued: "I would suggest that perhaps the best way to arrange this matter would be for you to phone me at your convenience." Cabot was not the only physician involved with such improvised arrangements. One patient wrote to him from New Hampshire in 1908 describing how a doctor whom he had visited in Boston "wished me to see other Drs before I returned home and he took me to two specialists [while in Boston]."[59] Anxieties about referrals may have encouraged this hybrid practice of visiting the office with the patient. These physicians were, after all, turning their private patients over to a new doctor. "He never came to me professionally after [seeing you], but told me on the short that you did not agree with me," one referring doctor wrote back to Cabot about their one-time mutual patient.[60]

Referrals were a flexible system that permitted the physician's services to be woven into a wide range of parallel professional and charitable services available to the sick. Cabot made use of referrals extensively to manage the many different problems that he identified as afflicting his patients. People who stayed in sanitariums for specialized treatments were both referred out and received back from their stays, like a woman from the neighboring town of Roxbury, Massachusetts, who was pleased to report back that after a fourteen-week stay in one facility, she "came home nice and well." Osteopaths referred patients to Cabot, as did homeopaths. Cabot in turn referred patients to an osteopath on occasion. He also used the group spiritual counseling by Elwood Worcester's controversial Emmanuel movement as a kind of referral, noting in the charts of patients, for example, that he had sent a patient on to Worcester for further care. He sent a destitute patient to Boston Consolidated Charities and a patient without a job to a restaurant where he expected that they might be hiring. These actions might reasonably be construed as a kind of referral system, and Cabot recorded them in his notes using the same format as for referrals to specialized physicians.[61]

Referrals also organized and permitted access to an ever-growing range of specialized physician's services, suggesting a seemingly limitless potential in modern medical progress. There was always something new, or at least something different, to try. Patients in Cabot's clinic followed fairly complex and exhausting paths on occasion through the clinics of his colleagues. Between 1902 and 1909, Mrs. Moore, the middle-aged woman whose path I followed through the wintery streets of Back Bay, saw Cabot sixteen times for related difficulties. She went to him first reporting recurring attacks of "buzzing" or "ringing" in her ears, along with dizziness, headaches, nausea, and sometimes vomiting. She described her troubles in her letters variously as "sick headaches," "biliousness," and "roaring ears." Her quest took her first, in 1902, on the referral of Cabot, to an otolaryngologist, who treated her without any great success. After trying several medicines to treat her himself, Cabot then noted that he had "referred [her] to Crockett," which meant Dr. Eugene Crockett, a second otolaryngologist. After four unsuccessful visits there, Cabot referred her next with worsening troubles to Dr. W. H. Kilburn, an ophthalmologist, who fitted her with eyeglasses. Dr. Kilburn wrote to thank Cabot "that you were good enough to refer to me" and to offer the modest claim that "if you find nothing besides eyestrain to account for her vertigo [dizziness], I shall be surprised if she does not get relief." For more than a year, she was putatively better with her glasses, but the dizziness returned after she had a hysterectomy in 1905. In the prior year, she had visited Cabot several times for heavy menstrual flow, and after treatments including "injection of citrate of iron," had a hysterectomy performed by a surgical colleague, and distant relative, of Cabot's. She returned to 190 Marlborough Street for postoperative care, and began to report dizziness again. For the next couple of years, she continued to seek relief in the offices of a third otolaryngologist, and again with Dr. Kilburn to adjust her eyeglasses. Cabot noted that these visits were sometimes "without any consid[erable] gain to her," but on other occasions, the attacks seemed "less severe and sudden." In July 1909, Cabot referred her back to Dr. Crockett to try his original treatments again.[62] She was never completely well, but she never stopped trying to get the right help.

The specialized services of Cabot and his colleagues must have derived part of their attraction from the promise of new medical science. To a public familiar with the remarkable stories of nineteenth-century rabies vaccines, diphtheria antitoxin, and, briefly at least, Dr. Koch's tuberculin, there was always the possibility that the next scientific breakthrough would provide assistance. Mrs. Moore gave her own testimony to the powers of this promise. By 1914, Cabot had communicated to her his strong pessimism about the prospect of finding a lasting medical solution for her troubles. But her own pessimism was more easily shaken. In February 1914, twelve years after she had started seeing Cabot, she wrote concerning an exciting new development. Dr. Crockett, she reported, after several other unsuccessful attempts, "wanted me to try a new treatment—radium at the Huntington Hospital." She reported that she had shared Cabot's discouraging prognosis with Crockett, only to have him dismiss it. When Crockett suggested to her that radium treatment might finally "stop the roaring in my ear . . . I told him you said it would never stop." But medical science provided an effective response. Dr. Crockett, she reported, had replied that "he would have said so too a short time ago," but then radium had appeared on the scene. There was a new cause for hope, a reason to try additional therapies and the doctors who provided them. Mrs. Moore went to the Huntington Hospital for radium, expressing a blend of anxiety and hope about specialized medical procedures that is common in the correspondence of Cabot's patients.

A system of referrals offered one means of endorsing the promise of new medical techniques and of channeling the aspirations that it stirred. People writing to Cabot seemed willing to try to create such a system where they found it lacking. In June 1909, a letter arrived in Cabot's office from a desperate father seeking advice on "my little boy 4 1/2 years old" who had become tragically sick. The trouble started when the child "became ill 3 weeks ago and 1 week ago the local physician diagnosed the trouble as leukemia and . . . [a specialist] confirmed this 5 days ago and classed it as lymphatic." This father wrote to Cabot, he explained, having found that he was the author of the chapter on blood diseases in a prominent medical textbook. The father communicated that he had the resources available to seek out the very best medical care for his child. His question was where to look. He explained the awful challenge facing him and his son: "As all authorities seem to agree the disease is necessarily fatal in the present state of medical knowledge my only hope is to get in touch with those who are close to the centers of research and study that if the little boy under the present treatment is temporarily restored we may learn quickly of any new discoveries in the treatment of this disease." Perhaps a new advance in science could save his son. Would Cabot refer him to the proper specialist?

Referrals could engage and support the hope for something better among patients. Mrs. Moore moved widely through the circles of Boston specialists as she pursued different evaluations and treatments. Nonetheless, she seemed to maintain a model of referral in which Cabot certified her quest as appropriate and perhaps offered reassurance that she was not straying too far. Considering the option of the radium, she wanted his advice, writing, "I longed to [come?] and ask you if I should [try the radium treatment]." Her solicitation of advice preserved the stability of her relationship with her physician, who had been directing her use of different specialized medical services for years. As she asserted about her own and her family's medical care: "[Y]ou see we can't bear to do anything without your approval." No records remain with the correspondence to indicate Cabot's response about the radium. But there would likely be other possible treatments ahead if the radium failed to help.

Not all who went searching for medical assistance availed themselves of the reassurances of a referral. Other patients sought their care independently among a complex array of practitioners without the guidance of physicians or other practitioners. Some of them found their way to 190 Marlborough Street. Among the people who reached Cabot's office independently, many arranged an appointment themselves, but others must simply have arrived at his door. One woman who had grown dissatisfied with her local physician had heard of him and decided that she would visit. She was the correspondent mentioned earlier who wrote asking if he would "please state what days and hours you are in your office and give me your address?"[63] She would presumably just drop in. People who had regular physicians in attendance could also, of course, arrange to see Cabot on their own initiative, without the knowledge of their physician. One woman who had recently visited wrote back recounting how "my doctor here dropped in the 14th and I told him I had been to see you. . . . He seemed surprised."[64] The chart of her original visit had not recorded the existence of this other physician. Dissatisfaction with service from another physician could be a reason to make an independent request for care from Cabot. A man writing to Cabot in 1912 about his father expressed an interest in leaving behind the physicians who were caring for him. "The local physicians in [his town] seem to me to be unsatisfactory," he wrote, adding, "at least they have not diagnosed his case and he has not obtained any relief."[65] Patients typically assumed that some preliminary arrangements were necessary for a visit, and many tried to set an appointment. By 1912, they might telephone to establish a time to come to the office.[66] People sometimes announced a special concern that triggered their need for medical attention, as with the woman who wrote, "I feel uneasy about the small lump that has formed in my right breast and would like your opinion before seeing any one else. Is it possible for me to make an appointment for Friday?"[67] She also suggested that Cabot's office would not be the only stop on her quest for assistance with this problem.

Conversely, of course, Cabot could visit his patients in their homes. Some people requested this, such as a woman in nearby Brookline who wrote in 1901 describing "nausea" and "gripping pains in my bowels-nothing serious-but very disagreeable," saying, "I would like to see you if you can pay me a visit."[68] He occasionally made home visits for established patients for a single treatment or assessment, but he rarely returned to orchestrate care for an entire episode of illness. Among the loose-leaf notes from his early practice records is a case from 1897 in which he had initially been called at 10:30 P.M. to the home of a patient who had fallen from a rocking chair. He subsequently provided a long series of home visits over the next several weeks attending to her broken arm. By 1900, however, cases of extended home visits had become very rare in Cabot's practice.[69] Home visits were seemingly unusual enough later for him to note one in the margin of the patient's chart. More commonly, when Cabot did visit a home, it was as a consultant meeting with the attending physician.[70]

Cabot's Clientele

An appreciable part of the analysis in this book derives from what Cabot's patients had to say about their health and their medical encounters. We might wonder, then, about their identities. Who made up this clientele? Where did they come from? What were their livelihoods and backgrounds? Cabot kept careful, but carefully circumscribed, records of the identities of his patients, which allow some conclusions about the people who constituted his practice.[71]

Cabot's private patients were in general neither Boston's wealthiest citizens nor its most disenfranchised. Many came to him in Back Bay from the circle of suburban towns that grew up around Boston in the late nine-teenth century. Not all patients have both an occupation and an address listed in their records, but those who do give a sense of the range of people who found their way to 190 Marlborough Street. Among patients whom Cabot saw in the first few years of his practice were, for example, a tailor from Boston, a professor from Salem, a carpenter from Watertown, a nurse from Jamaica Plains, a garment inspector from Somerville, a masonry boss from Everett, a piano tuner from Dorchester, and a law stu-dent from Cambridge.[72] The occupational backgrounds of his patients were clustered among professionals, merchants, and skilled laborers, but he also exceptionally saw the son of a prominent industrial tycoon and an unemployed immigrant.

The leveling effects of office medical practice also shaped the presen-tation of the data on patients' identities. Their occupations provide some clues. Although patients in the sample were aged from four to eighty-five, most were working adults, with an average age in the sample of thirty-eight and a median age of thirty-nine. In the random sample, slightly less than one-fifth of the women have an identifiable occupation noted in their charts, compared to three-fifths of the men. Many of these women are listed as married and likely worked exclusively in their homes. But I found nothing to reassure me that there were not many whose other occupations Cabot systematically neglected to inquire after or note. The occupations listed for the 104 women in the random sample range from kitchen girl, chambermaid, and former nanny to bookkeeper, nurse, and saleswoman. It is noteworthy that 5 out of the 104 women in the sample have their hus-bands' occupations recorded as their own, even when the husband's liveli-hood was no longer directly relevant (e.g., "junk dealer's widow"). Was the woman recorded as "salesman" married to a salesperson or one herself?[73] Work limited to the home was typical for wealthier married woman in the early twentieth century; but perhaps Cabot's assumptions about their roles led him to miss other information. After all, a practice in an urban medical office may well have selected for an independent group of women inclined to venture outside of their homes for care.

Judging the occupations of male patients entails a different kind of bias. Cabot tended to flatten out the reported occupational status of his patients. A range of occupations is represented, from lawyer, judge, profes-sor, medical student, and "flannel manufacturer" to "peddling fruit," cab driver, stable keeper, and steel laborer. The occupational records are, in addition, skewed toward the professional by a number of physicians who were patients. In the random sample of two hundred patients, there were six male doctors and three women recorded as the wives of doctors. Given the nature of these records, however, someone whose occupation is recorded as "watch factory" may have worked in a factory or have owned it. These records tend to blur the distinctions between very different occupations, emphasizing the product of the work rather than the position of the worker. So while one person identified as a "shirt waist maker" actually did sew shirtwaists, based on the information in the notes, another man described only as a "piano-maker" is identified in an accompanying obituary as the treasurer of a prominent piano factory. A man recorded only as a "roofer" is shown by his own impressive letterhead to have been the owner of a commercial roofing company, while a man whose occupation is recorded as "iron foundry" is described in a letter from his son as the foreman of the foundry. The occupation of the man called a "janitor" may seem clear, but someone whose occupation was "shoe factory" may equally well have been the owner, the floor supervisor, or a day laborer. The instance of a man whose work is noted only as "saloon" raises interesting questions. Did a man of whom it was noted, "sells candy," sell it by the bag from a stall, one wonders, or by the truckload from his factory?[74]

It is tempting to speculate on the significance of Cabot's flattened occupational designations. He tended to level the socioeconomic characteristics of patients, so that in these medical records, the piano maker and the shirtwaist maker seem more similar than they would have appeared outside the medical setting. The important differences between patients, the differences highlighted by these records, are differences in medical condition or diagnosis. Medical categories superseded other social roles. Occupations received particular attention, for example, with regard to their specific effects on health. Thus, the occupation of a patient who was pale and weak might be recorded simply as "indoor work," capturing tersely a significant medical feature of a job away from sunlight and fresh air. With other patients, the nature of their occupations seemed more significant, and Cabot specifies, for example, that one man's occupation was "scouring wet cloth finishing (10 yrs)."[75]

Gender rather remarkably offers a similar example of the leveling effects of the medical condition in certain instances. Roughly half of the patients visiting this office were women, 52 percent in the random sample. Typically, it was easy to discern the gender of the patients from these records; yet in some cases it was difficult and in others impossible to do so. These records classify people primarily according to disease, and only secondarily according to more commonplace social identities, like gender, age, occupation, or race and ethnicity. In the cases of a patient with pernicious anemia and another with myocardial weakness, for example, the nature of the disease is evident, but not the patient's sex. Listed only by their surnames, without other identifying information, these patients exist in the charts only as diseases.[76] Pernicious anemia meant special blood testing and treatments with arsenic, x-rays, thorium, or splenectomy. Myocardial weakness might entail advice about a regimen of rest and the use of stimulating medications like digitalis. The identity of these patients beyond their need for specially defined medical services is obscure.

The strongest example of the suppression of social identity is in the instance of race. Scattered observations on the racial and ethnic characteristics of Cabot's patients reveal more by what they omit than by what they include. Cabot identifies a national origin for very few of his office patients, noting occasionally that the individual was Swedish, or from "Italy." A couple of patients are noted as "colored." Another small group of patients are identified as "Hebraic"—sometimes derogatively.[77] Most patients, however, received no ethnic or racial designation. The great majority in this group with Anglo-Saxon surnames, like Miller, Burt, Greene, Taylor, Webster, Allen, or Drew, are not otherwise identified in terms of ancestry or ethnicity. This suggests that the group warranted no separate identification in Cabot's view.[78] Still, the names of the patients indicate a diversity of backgrounds that found no other direct expression in Cabot's clinical records. It is noteworthy that one significant group among Boston's citizens is almost unrepresented among the clientele of this clinic. Judging by the surnames in these records, Boston's Irish seem not to have gone to 190 Marlborough Street for their medical care, although they would have been well represented among the patients that Cabot saw in the Out-Patient Department at Massachusetts General Hospital, just across town.[79]

Race mattered to Cabot in clinical reasoning, as he explains in his textbook on diagnosis in various places. Referring to specific cases, he says, for example, that "any pain of this type occurring in a man of fifty-nine suggests aneurysm or angina pectoris, especially if the patient is a negro," and that "there is no reason for accusing the stolid Italian laborer of the 'vapors.' " although he does not elaborate on the basis for these observations.[80] People of different racial and national backgrounds suffered from different distributions of disease and described their symptoms in what seemed predictably different manners. But once again the significance of disease seems to have overwhelmed the need to account explicitly for race in these records. The fact that it appears in the records prominently only in the case of nervous disorders is explored in chapter 5 below.

Sliding Scale of Payment

Cabot's medical records configured the patient's social role as an element of information with specific relevance to medical practice. Medical authority in the clinic functioned partly through the ability to create these transitions and bridges from the patient's external social world into the oddly intimate, technical focus of the clinic. Social roles and identities mattered, but they mattered most in their implications for medical management. Setting fees for medical service offered an interesting expression of the power to translate occupation into information relevant to practice. A so-called sliding scale of fees linked the patient's perceived socioeconomic status to the amount charged for service. Cabot used such a sliding scale of fees, and the letters of other physicians and patients suggest widespread familiarity with this system and general anticipation of its use. Patients who seemed able to afford higher fees were in general billed more, while those who seemed poor were billed less. While referring physicians took the most visible role in negotiating fees, patients too petitioned for reduced charges based on their means, circumstances, and occupation. They occasionally sought a lenient fee without any reference to their means, as in the case of a patient who wrote on her personal stationary in a tidy hand: "If it would not be trespassing too much on your kindness and time would you let me come to you for the same sum ($3.00) you allowed me to see you for eight months ago?" But more typically, people appealed to the doctor on the basis of their work, salaries, and financial obligations.[81]

The referring physicians usually petitioned on behalf of their patients, as when Dr. Chase sent a patient bearing a letter to Cabot describing her care and noting that "their circumstances are moderate. Husband is a grocer."[82] In other cases, a more complex set of considerations was invoked. Writing in 1905 that her patient's husband was of limited means and "if in justice to yourself you could show him any special consideration in the way of a slight reduction of fee, I assure you I would appreciate your kindness," Dr. Sarah Bond seems to petition as much on the basis of her relationship with Cabot as on behalf of her patient. This patient's husband, she explained, had a reasonable salary as a university professor but was "unfortunately carrying a heavy load of both his wife's family and his own."[83] Another physician wrote, similarly, "What I wanted to do was to see if I could make some arrangements . . . that the cost to him would be as little as would be consistent with his means."[84] Other subtleties of negotiation are also evident in the records. One physician wrote of his patient to Cabot that "he is not in very good circumstances but would raise the money somewhere if he had to go and see you." When he saw this patient on referral, however, Cabot noted in his chart that this man had already "been osteopathic—$90 worth." Apparently, the patient in question found money to pay other practitioners. Cabot submitted and was paid a bill for $15 for the visit, not substantially less than similar charges in the same period of practice.[85]

The sliding scale went up as well as down, although discussions over higher fees were considerably more circumspect. The physician's willingness to lower the expected fee in the case of straitened circumstances was naturally a more acceptable message to convey. There were few people who wrote, as one did, saying, "as I am not one who wants 'something for nothing' I hope you will charge me well for all your trouble," but colleagues of Cabot's seem to have been willing to hint when someone could well afford the fees.[86] Some physicians writing in referral seem to imply that they are seeking not a reduction in fees from Cabot but restraint in increasing them, as in the case of the doctor who wrote to ask Cabot to "kindly make the charge as reasonable as possible," explaining, "patient not in affluent circumstances."[87]

More went into Cabot's assessment of his fees than the patient's ability to pay, of course. The most important factor was probably the nature of the services rendered, although this is often difficult to assess from these records. A statistical analysis comparing the fees charged in individual cases with the number of laboratory tests performed suggests a correlation here.[88] Consultations were more expensive on the whole than referral visits, accounting for the costs of Cabot's travel. In addition, the vagaries of Cabot's reporting on occupational status make it hard to compare occupations with levels of charge. On three separate visits during one week in 1914, Cabot billed a singer nothing, a carpenter $15, and a "telephone man" $25. The next week a "judge" had a visit, recorded in some detail, for which he was billed $5. The effects of the sliding scale are most evident at the extremes, with patients who were recent immigrants paying unusually low fees, while a wealthy tycoon received a bill for $200, which he seemingly paid without hesitation.[89]

Cabot occasionally provided free services to patients. In some cases, it was simply charity. But free services functioned in other ways too. In 1913, Cabot saw an insurance agent suffering from "bladder trouble" and advised him to consult a urologist. No fee was charged for the visit, with the notation "free," added in the chart.[90] It is possible that this man was actually unable to pay, but his visit seems to have been a brief one, judging from the records, and perhaps Cabot felt that he had performed only a minimal service. There was likewise no charge, perhaps on similar grounds, when Cabot saw a child suffering from bedwetting and found that "all my measures tried and n[o] g[ood]." It is equally possible that other factors influenced the provision of free care. Cabot generally offered professional courtesy to other physicians, to their family members, and to nurses, for example. Other personal connections counted as well. Cabot also provided free services to various of his relatives when he saw them as patients.[91]

THE SLIDING SCALE OF MEDICAL FEES linked the provision of care with the patient's economic and occupational standing and softened, or at least obscured, the influence of social position or wealth in controlling access to care. At least figuratively, it took money off the table in negotiating services with a doctor. In some markets, you could buy whatever you could afford. A sliding scale made obtaining a doctor's services depend less on what you could afford and more on what the physician was willing to provide. The sliding scale provided at least some potential shelter from competitive market pressures, and so expressed the interests of professional autonomy as much as the obvious desire to make a good living. Perhaps a wealthier client could obtain better services merely by paying more—but perhaps he only paid more for the same services.

The issue of the sliding scale provoked lively debate among Cabot's peers in the early twentieth century. Doctors of the day typically offered two slightly different justifications for the sliding scale, depending on whether they discussed only lowering fees or both lowering and raising them. The justification for lower fees was the obvious argument for charity. Physicians arguing a disinterested case for the raising and lowering of fees had to appeal instead to the general social good. The appeal to charity seems to have been more common and was more readily defensible.[92] Dr. Roger Lee, also of Boston, argued that charity allowed physicians to reduce their fees to accommodate a limited income. They should not, however, raise fees just because other patients were able to afford them.[93] Richard Cabot's brother Hugh made one of the more cogent arguments based on justice, contending that sliding scales should run both ways, so that wealthier patients subsidized the care of poorer ones. He went so far as to compare the sliding scale of fees to a system of graduated income taxes. The sliding fee permitted a measure of equity, administered by the physician.[94]

The discussion here will leave aside the question of how well or how fairly the sliding scale worked, which would be difficult to resolve given the nature of the evidence. Critics have been quick to point out that sliding scales provided a handy mechanism for maximizing medical fees without much oversight or disclosure.[95] Physicians charged whatever they thought a patient could pay and claimed to be obliged to do so by an open, competitive market. In the face of criticism over fees, turn-of-the-century physicians tended to fall back on the general defense that however they billed, they frequently went unpaid in any case; and patients, they noted, tended to create their own sliding scales, paying whatever a service seemed to be worth to them.[96] Stories abounded in professional communications of physicians who were able to exact a fair payment from a reluctant patient, usually through gentle subterfuge. But such stories most often took the form of jokes.[97] Doctors took care not to seem too serious about this matter, perhaps reflecting the delicate task of preserving the image of beneficence when they routinely earned their livings from families in circumstances of crisis, debility, sickness, and loss.

Whatever its actual use, the sliding scale offered American physicians at least hypothetical shelter from external economic pressures, sheltering physicians in part from a competitive market, while equally seeming to buffer the individual physician against the influences of wealthy patronage. Such arguments appealed to Richard Cabot, with his concern for the physician's autonomous and legitimate control of practice. Cabot echoed a common criticism by American physicians before the advent of medical insurance in arguing that the best medical care went to the wealthiest, who could afford the full cost for surgeries, x-rays, or hospitalization, and to the poor, who often received identical, charitable care. It was the middle class, Cabot argued, who were shut out from the most advanced medical services. Only a system of group, prepaid medical insurance, he claimed in his popular writings, would remedy the flaws in a system of direct payment for services that relied on medical charity and covert shifting of costs.[98]

CHAPTER 3

THE DIAGNOSIS OF HIDDEN DISEASE

The work of diagnosis in early twentieth-century medicine posed a challenge to academic physicians like Cabot. A case from his clinic helps to highlight the kinds of problems that could arise. In 1939, a man wrote to introduce himself as a former patient, looking to reclaim the benefits of earlier medical attention. He described in his letter how in May 1919 he had gone to Cabot's office on Marlborough Street with an undiagnosed problem. "No doctor knew what it was," he recalled. It was only after Cabot "made arrangements" for him to go to Massachusetts General Hospital that the physicians there "through their daily examinations, blood tests, x-ray pictures etc. were able to discover the cause and gave me a medicine that eventually cured me completely."[1] He was writing now, he explained, two decades later in the hope that he could get some more of that original medicine. He felt his old problem was returning.

Adding this letter to his records Cabot may have noticed a familiar irony. The man recounted the story of a specific ailment, uncovered by a vigorous diagnostic investigation. He believed that the "cause" of his problem had been found and successfully treated. The physicians who examined him at Massachusetts General in 1919, however, provided a very different account in their report, filed alongside Cabot's original notes. They had written to Cabot in 1919 to explain that their long course of testing regrettably turned up "no diagnosis beyond chronic inflammation" and that they prescribed for this man only a general tonic containing a sedative medication.[2] Physicians and patient differed considerably in their understanding of the diagnostic process and its outcome in this case. The struggles of Cabot, his patients, and his colleagues to manage the process of diagnostic evaluation tell us a good deal about the expression of individual medical authority.

Differences between what physicians and patients understood about diagnosis and disease were certainly not new to the early twentieth century. Yet such differences bore new significance as they aligned with a gap opening between the rigorous, and increasingly technical, pursuit of specific disease and the reception of this process by patients.[3] Dr. James Herrick, addressing a medical audience in 1928 on the previous decades' innovations in diagnostic bacteriology, pathology, chemistry, and radiology, told his colleagues that in the twentieth century, the process of diagnosis had become, "while far more accurate, much more difficult than it used to be."[4] Medical science provided a growing array of new tools for isolating and identifying human diseases. This power to define a disease in the laboratory or the medical textbook did not, however, translate directly into the capacity to diagnose it persuasively in a patient, as turn-of-the-century medical essayists sometimes noted.[5] Cabot's colleagues at Massachusetts General in 1919 had been unable or unwilling to explain their ambiguous diagnostic results to the patient-perhaps being themselves unsure whether they had reached no diagnosis or had diagnosed the absence of disease.

A consideration of diagnostic practice among Cabot and his colleagues speaks to larger questions about the emerging constitution of the physician's individual influence in this period. The confidence manifested by Cabot and his peers in the technical identification of disease would not necessarily command the support of patients in face-to-face encounters. Twentieth-century physicians increasingly drew upon a reputation for technical expertise to bolster their persuasiveness, but they had to deploy this expertise carefully in private practice. Cabot was a leader in an early twentieth-century program to perfect diagnostic reasoning skills. His application of these ideal methods and processes in his clinic were particularly revealing of the inherent tensions involved in making diagnosis work in daily interchanges with patients. Cabot's innovative Clinicopathological Conferences made a compelling display of the doctor's idealized power to diagnose hidden disease and became a widely emulated model for diagnostic practice. The patients whose cases were presented at these professional exercises were, however, never present at them. How they might have responded in person is suggested by Cabot's clinical records, which document negotiations between doctor and patient, with diagnostic strategies, data, and opinions recorded in the charts and patients engaging with them and interpreting them back in their letters.

Influence and Obligation in the Diagnostic Process

Cabot shared with some of his colleagues a concern about how far a doctor should, or could, go in diagnosing a specific disease. What might be sacrificed, for example, in striving for a definitive diagnosis? As the tools of radiology, chemistry, surgical histology, and bacteriology steadily solidified a claim for use, physicians advised different strategies for managing their patients through a prolonged and complex diagnostic process. It remained unclear how the conduct of diagnosis would affect their obligations to their patients and their influence over their patients' choices and responses. A question emerged in the medical literature of the period about the kind of medical authority that was defined and expressed in these technical practices.

One of the more outspoken participants in this debate was Alfred Worcester, a busy private practitioner, professor of hygiene, and an influential member of the Massachusetts Medical Society. A senior colleague of Cabot's in Boston, Worcester was an unusually self-assured critic of the harsh and distorting effects of a rigorous diagnostic practice. In an intriguing essay published in the *Boston Medical and Surgical Journal* in 1912, Worcester protested against what he saw as medicine's excessive enthusiasm for lengthy evaluations, which distracted physicians from other humane obligations. "The whole strength of the [medical] schools," he noted, was "devoted to the study of diseases," without training physicians sufficiently in the skills of cultivating personal influence. "In general practice," he wrote, "poor diagnosticians are often very successful because of their knowledge of human nature." Worcester went on to explain what he meant by "successful," illustrating how the demands of modern diagnosis diverted doctors away from developing and exercising their traditional "knowledge of human nature."[6]

In this essay, Worcester used a series of hypothetical scenarios, or medical parables, to contrast a traditional form of practice with the methods of the modern physician who pushed diagnostic ideals to unnecessary lengths. In his first scenario, Worcester sketched a sentimental portrait of an old-fashioned doctor visiting a woman caring for her desperately ill child. Diagnosis is far from this traditional practitioner's mind. He "does not thump or even stethoscope the [child's] wasted chest," relying instead on a general sense of what will be beneficial, drawn from his long experience with the family. The traditional doctor, Worcester argued, had "no need of a scientific diagnosis or prognosis," although he easily perceived the child's grave condition.[7] Unconcerned with specific questions about the nature of the culprit disease, he devotes himself instead to supporting the mother's attentive care of her child by encouraging additional feedings and more fresh air.

Worcester then depicted an enthusiastic, modern diagnostician of 1912 moving through the very same scenario, with different goals and results. This modern doctor examines the same child's chest intently "to percuss it, and to listen through his stethoscope." Identifying a life-threatening disease, he dutifully tells the mother. The diagnosis affects her badly. She struggles to continue providing care for her child, but turns away from the physician, seeking someone who will "at least leave some hope in the sick room."[8] Worcester expressed no particular criticism of the tools and techniques required here for diagnosis. What seemed to trouble him was how the requirement of making a diagnosis altered the doctor's interactions with patients. The doctor was so deeply absorbed in the immediate technical goals that he sacrificed an ongoing therapeutic influence over his patient. Worcester reportedly claimed proudly that his own patients often got well "without the luxury of diagnosis."[9] He seems to have been following the strategy pursued by the old-fashioned doctor of his parable.

The diagnostic process, as Worcester characterized it, was compelling and persuasive, but perhaps too much so. It distracted physician and patient alike and threatened to undermine the healing influences that might otherwise be wielded by the doctor. In another scenario, Worcester went on to outline the equally detrimental effects of exact diagnosis even in less serious diseases. He caricatured his same modern diagnostician facing a fussy older woman who presents a complicated story of many discomforts and maladies. Worcester's disease-oriented doctor feels obliged to pursue a "minute examination" of the woman. Unfortunately, this enthusiastic diagnostician, who excels "wherever the knowledge of diseases is needed," does not understand "that persistent search for the abnormal in human bodies will reveal some trifling trouble nine times out of ten." Such a physician, Worcester argued, was apt to make this patient feel worse. The woman in the example would become more anxious about her health under his scrutiny, and so less healthy. Once again, the traditional doctor appears as the foil. He is portrayed as benefiting this same patient by leaving out most of his examination and diagnosis. Without denying to this older woman the possibility of some mild disorder, he uses his personal knowledge of her family, neighbors, and circumstances to contrive an effective treatment for her. Worcester ends his story with the traditional doctor cleverly turning this woman's attention away from her own problems to helping others. This sort of medical influence, Worcester suggested, would prove more helpful than the exacting pursuit of a diagnosis.[10]

By 1912, Alfred Worcester must have been fighting something of a rearguard defense in rejecting the need to "thump or even stethoscope" his patient. A succeeding generation of physicians around Boston already found a benign authority evident in the workings of sophisticated diagnostic processes. In 1927, Dr. Francis Peabody, a later contemporary of Cabot's in Boston's medical world, published an insightful and still much admired essay on "The Care of the Patient." Just a year earlier, Peabody had himself received from his doctors the diagnosis of a fatal cancer, after a prolonged illness and a lengthy diagnostic investigation. During his lingering illness, he developed a set of moving essays on what he understood about the humane obligations of physicians to their patients. The writings addressed several obstacles that he felt faced young physicians and medical students in learning to provide humane care. The process of diagnosis especially drew his attention. He portrayed an extended process of diagnosis as fundamentally beneficial, if properly conducted. The problems, in Peabody's view, arose when there was insufficient personal oversight of the process or premature conclusiveness. Detrimental effects typically arose not from too much attention to diagnosis, but from too little. Peabody warned the novice physician, for example, that "[a]t its worst, however, the Diagnostic Clinic is a machine, and the patient is automatically passed from one specialist to another, and submitted to a series of examinations." Such a process was potentially hazardous for the patient, in part because of "the lack of some one man who understands the situation as a whole."[11] He exhorted young doctors that diagnosis required careful supervision by a single physician who would take responsibility for the process and guide a patient through the interpretation of its conclusions.

With proper oversight, an extended course of diagnostic evaluation actually cemented a close relationship between a doctor and a patient, Peabody suggested, thereby creating grounds for continuing medical care. In arguing this case in "The Care of the Patient," Peabody sketched a scenario similar to the one that had worried Alfred Worcester. A man is admitted to the hospital and tolerates a week of lengthy diagnostic examinations, which turn up absolutely no evidence of any disease. Doctors might feel that they could then discharge their duty to a patient by explaining simply that "there's nothing the matter." This would be a mistake, Peabody advised. Physicians should remain cautious about reassuring their patients, or themselves, about the complete absence of disease. The doctor must, for example, consider whether the symptoms "are the result of an organic disease in such an early stage that you cannot definitely recognize it."[12] Perhaps future examinations would reveal the disease as it developed—as had been the case with Peabody's own fatal condition, in fact. Symptoms could be monitored and reevaluated. This indefinitely long process of diagnosis, even if it did not result in the identification of a disease, still proved helpful in defining and supporting the relationship between doctor and patient. Both the reassuringly negative results of the testing and the physician's demonstrated commitment to finding any disease that emerged were ultimately beneficial in themselves, in Peabody's view.

The Clinicopathological Conference

In his clinic, Cabot faced scenarios loosely similar to those sketched by Worcester and Peabody; but his diagnostic practices differed sharply from what these two doctors prescribed. The daily records and correspondence of Cabot's private clinic captured a more complex portrait of diagnosis than a rhetorical essay, of course. But in addition Cabot seems to have had a different conception of the authority properly expressed in the diagnostic process. For Francis Peabody, diagnostic vigilance was one way for physicians to demonstrate their benevolent attentiveness to a patient, even without reaching a diagnosis. Alfred Worcester argued instead that the hunt for hidden diseases might distract doctors and loosen their hold on traditional forms of medical authority, exercised through personal moral influence. Cabot, however, shaped a model of diagnosis that emphasized the impersonal, objective authority of the doctor's diagnostic opinion. Diagnosis should be neither avoided nor postponed. In his practice, he required a diagnosis as a necessary first step, preceding and justifying medical advice or intervention. For Cabot, expert knowledge about disease formed the legitimate basis for medical influence and control.

It is necessary to pause here to note that readers who are familiar with Cabot's famous reforms in medical social work and pastoral counseling may find it puzzling that he adhered to such a strict disease model in diagnostic practice. Cabot was known in his day, and is remembered even today, for championing attention to the social and spiritual elements of the sick person's life. Although in diagnosis he emphasized a narrow vision of specific diseases, he argued in numerous publications that the range of problems to be addressed properly in illness was broad and inclusive. In a list that he compiled of the major scientific advances in medicine between 1900 and 1910, Cabot included the use of "antitoxin for epidemic meningitis," the discovery of x-ray photography, and the recognition and care of "mental and moral factors in disease."[13] Sounding at times more like his colleague Alfred Worcester, he noted that his medical peers were unfortunately beginning to drop their traditionally broad attention to the many factors that contributed to a person's illness and health. Cabot protested that in busy, crowded modern clinics, "our patients shoot by us like comets, crossing for a moment our field of vision." The elaborate technical resources of medicine encouraged the busy doctor "to focus upon a single suspected organ till he thinks of his patients almost like disembodied diseases," with the result that less time was devoted to the complicated circumstances of an individual patient's life.[14]

Cabot did not, however, extend his critique to arguing that individual physicians should routinely pursue the personal, social, and spiritual dimensions of healing. Instead, he lobbied to delegate responsibilities for these components of illness to other professions. His efforts to involve social workers, counselors, and clergy with medical care were premised in part on an understanding that physicians were typically already fully occupied with their responsibilities for disease. Although he lamented fading attention to "*personal* service of the simplest kind" and speculated on the potentially harmful effects of this narrowness on the character of the average physician, he did not argue that this problem should be addressed by changing the physician's vision and approach. Cabot did not, for example, lobby to train physicians in social work or pastoral counseling. Like many of his progressive-minded colleagues, he proposed that the segmentation of tasks, the training of expertise, and the creation of integrated institutional structures would solve such problems. He worked to integrate the technically accomplished physician into a system of specialized professional workers who would collaboratively cover the various areas of need for afflicted patients. It was a solution that appealed to a progressive reformer like Cabot, who saw a solution to the chaos of modern, urban life in the power of organized expertise.[15]

To initiate his proposed reforms, Cabot founded the first American hospital-based department of social work at Massachusetts General Hospital in 1905. Over the next two decades, in articles, books, and talks, he widely promoted the virtues of professional medical social workers. Treatment of disease was not the entire solution to health. He envisioned a "rebirth of therapeutics," with conscious and systematic attention to the traditional nineteenth-century elements of diet, occupation, environment, and the moral and emotional status of the patient—the core responsibilities of the physician in Worcester's telling. In Cabot's scheme, however, it was the nurse, the clerical counselor, and the social worker who would take responsibility for these dimensions of care. The technically skilled physician, who was necessarily occupied elsewhere, would "rather turn [these details] over to some one else who is interested in them," as he put it bluntly on occasion. Other specialized professionals would help to reclaim access to the individual features of family, occupation, religion, and neighborhood that were being lost to the busy physician in the modern clinic.[16]

These reforms squared neatly with Cabot's very successful model for an exacting diagnostic process, established in the Clinicopathological Conferences that he founded at Massachusetts General Hospital, perhaps his most lasting and significant contribution to American medicine. These conferences enacted an ideal diagnostic process that is still routinely put on display today at MGH and in other hospitals across the nation, offering a persuasive demonstration of the physician's capacity to identify hidden diseases. Modifying some preexisting diagnostic teaching exercises, Cabot began scheduled performances of his Clinicopathological Conferences about 1909 at the hospital, where he had been recently appointed visiting physician. Within six years, Cabot began to circulate formal, printed accounts of the conferences to colleagues. As the exercises developed, they attracted a steady following outside of MGH, and they were selected for routine publication in 1924 by the *Boston Medical and Surgical Journal;* they still appear today, in roughly the same format, in the same publication, since renamed the *New England Journal of Medicine.*[17]

Diagnostic exercises had established a long tradition in clinical medicine by the time of Cabot's training. His own experience as a medical student in a similar exercise may have provided him with an early model for the Clinicopathological Conferences. In the 1890s, Professor Frederick Shattuck regularly conducted a diagnostic teaching conference with the medical students at Harvard Medical School. In these conferences, a student selected from the class listened to the information gathered from a diagnostic evaluation of a patient and then speculated before the assembled group about the underlying disease. Shattuck typically provided the final answer by reviewing the same diagnostic information and reaching his own conclusion, perhaps disagreeing with the student. The exercises had a pedagogical justification as an opportunity for Shattuck to review the student's reasoning about the diagnosis. At one conference in 1891, when Cabot was the featured student diagnostician, Shattuck had additional information from an autopsy performed after the death of the patient, which provided a definitive diagnosis, and Cabot's diagnosis was confirmed by the autopsy data.[18] Being a winner might, of course, inspire the sense that this was an excellent form of competition.

Years later, Cabot preserved a similar element of drama in his Clinicopathological Conferences, which began with the featured physician, often Cabot himself in the beginning, listening to a summary of diagnostic information collected from a single hospitalized patient, a boy with severe backache in the first published case in 1924. Following a description of the patient's history of symptoms were observations from the physical examination and the results of laboratory and radiological testing. One critical piece of information was, however, withheld from the diagnosing physician. The outcome of an autopsy or a surgery that had identified the active disease was kept hidden. Ignorant of the final answer, the featured physician reviewed the evidence and created a "differential diagnosis" listing the possible diseases that could explain the case. In the first published case, for example, these included causes of chronic illness and severe back pain like "tuberculosis of the spine," "staphylococcus abscess of the kidney," and "staphylococcus abscess of the spine." The featured physician then com-

pleted the first part of the exercise by venturing a definitive diagnosis. In
the dramatic conclusion, the examining pathologist stepped forward and
announced the actual microscopic and physical findings on autopsy, identifying the hidden disease. In the case of the boy with the back pain, the
participating doctor was told that he had correctly diagnosed the boy's disease as a bacterial infection of the heart, despite distracting secondary
involvement of the spine.[19]

Cabot's Clinicopathological Conferences offered a striking validation
of the physician's control over the process of diagnosis, demonstrating
how diagnosis had a basis in the systematic, objective, and often conclusive
search for hidden disease, which was publicly tested in this forum. Physicians in the conference were called on to reason out their diagnosis openly,
justifying their conclusions according to the information from the diagnostic evaluation. The conferences were repeated weekly in the hospital
under Cabot's direction, bringing home the same message each time.
There was a disease behind each patient's troubles. It was sometimes
deceptive in its manifestations, so that a pain in the back might originate
from an infection in the heart. But a physician would ideally reach the correct diagnosis by organizing the information properly and following it
attentively. These conferences offered a novel and dramatic endorsement
of the power of diagnostic evaluation and of the physician's legitimate control of the process, at a time when the technical resources for such evaluations were rapidly expanding.

One attraction of the conferences, and perhaps a source of their
longevity and success, was the way that they acknowledged and accommodated an individual physician's diagnostic errors within the process. The
format of the exercises required that each featured physician provide a differential diagnosis listing as many competing diagnoses as possible before
venturing a final opinion.[20] Uncertainty about the hidden disease and
acknowledgment of the potential for error in identifying it became formal,
predetermined parts of the conference. Cabot himself emphasized how
doctors could sharpen their diagnostic skills through repeatedly testing
themselves against, and perhaps failing to meet, an exacting standard like
the evidence from an autopsy.[21] Acknowledgment of the possibility for
error in the proceedings made a strong impression on observers, especially
those outside the medical profession. One social worker who routinely
attended the early conferences still recalled years later that "to see the doc-

tors accept the fact that they were wrong was a deeply humbling experience for the audience"—and for the doctors too, no doubt.[22] The Clinicopathological Conferences provided a vivid demonstration of medicine's responsibilities for the identification of disease, structuring individual fallibility as an anticipated part of a professionally sanctioned and supervised process.

Making a Diagnosis in the Clinic

Richard Cabot adopted the same ideal of diagnosis into his private practice. Yet, as Worcester and Peabody suggested, there was considerable flexibility in how the elaborate diagnostic process would fit into daily clinical work. Cabot attempted to follow a simple model. He set down the pertinent information, gathered in turn from the patient's accounts, from his own examination, and from the laboratory testing. Then, for each new patient, he proposed a diagnosis or, rarely, a couple of diagnoses, recorded almost invariably in the last lines of his first note on each case. On the page, the process that was described seemed like a brief version of the Clinicopathological Conference. These meticulous office records show Cabot collecting diagnostic evidence, winnowing out the possibilities, and staking his opinion. The diagnosis listed at the end of each case provided an explicit basis for subsequent medical advice and attention. Whatever influence Cabot wielded over his patients, it would have a documented justification in his claim about disease. Disease was the basis for his advice and recommendations to his patients, as represented in the charts.[23]

But diagnosis functioned less smoothly in the private office among patients than in the settled context of the hospital conference hall. The authority on display in an independent professional exercise differed from the authority endorsed or contested by the people who came to Cabot seeking to know what was wrong with them. In the Clinicopathological Conferences, the physician pursued the independent consideration of the evidence of disease generated in the diagnostic workup. Cabot's charts reflected this ideal, presenting the diagnostic process as a step-by-step pursuit of the evidence of disease that went on independently of the opinions and interests of the patient being diagnosed. But diagnosis in the active world of the clinic was a dynamic and reciprocal process. The people who came to Cabot spoke at length in their letters about their diagnoses and inquired after them quite eagerly. Their letters and notes also revealed

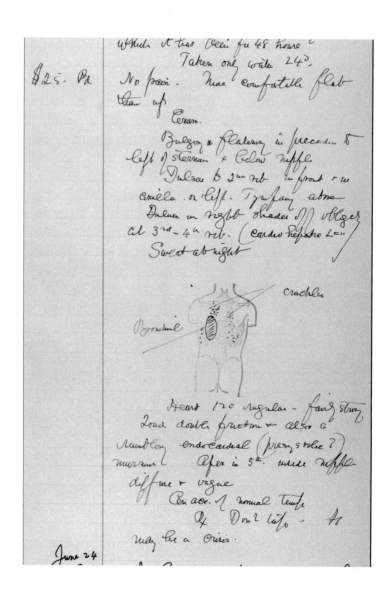

A representative page from Cabot's patient records, showing the use of an ink stamp of the torso to indicate areas of abnormality. Cabot's prescription here, referring to draining fluid from around the lung, is "Don't tap. It may be a crisis." Courtesy of the Harvard University Archives.

Volumes 7 through 12 of Richard Cabot's bound patient records. Courtesy of the Harvard University Archives.

complicated negotiations with Cabot, for which there was little room in the ideal model expressed in the conferences. Cabot used a standard format for his office charts, also widely used in other medical charts in the time, which presented the work of diagnosis as internal to medical practice and unreciprocated.[24] Patients writing back to him opened up a perspective on the mutually negotiated authority actually at play in the clinic.

Two striking cases from Cabot's practice records serve to illustrate the unexpected sources of support and challenge for the seemingly indepen-dent diagnostic process modeled in the Clinicopathological Conferences. A patient could, for example, reject Cabot's diagnostic opinion but nonetheless support the general principles on which it was based. One especially articulate, if quirky, sixty-five-year-old man from Dorchester, Massachusetts, wrote to disagree with Cabot's diagnosis but to endorse the idea that a rational, empirical search for disease was crucial. He had originally visited the office in 1902 and wrote back six years later to instruct Cabot about his and the medical profession's errors. Cabot and other doctors had been misdiagnosing him, he said. He had, however, been conducting his own inquiries and reported in his letter that he had "demonstrated at least to my own satisfaction . . . [that] all of my symptoms are mainly resultant upon the activity of that incurable and distressing disease called catarrh." He explained that he had confirmed the presence of catarrh and had treated it through his own "research, investigation and persistent testing."[25] His pursuit of the diagnosis seemed similar to his doctor's, only it had been more accurate by his account.

Other people disagreed with Cabot on the identity of their disease, such as the man who wrote politely suggesting that if his doctors were "not so sure of the cause and complaint, I should swear it was caused by a severe attack of malaria." Such correspondents rarely delved into the basis of their disagreements. The patient with the "catarrh," however, cited his familiarity with the limits of medical knowledge in support of his conclusions. He continued in his letter explaining: "None of the regular medical practitioners whom I have consulted in respect to my case will admit that catarrh is particularly accountable for my condition. . . . I fancy that they assume this attitude . . . because the science of pathology and therapeutics as taught as bookish theoric [sic] does not admit of their doing otherwise."[26]

Some patients felt they could challenge Cabot on his own terrain. A seventy-year-old man, described in the records as a former farmer, received the unacceptable diagnosis of a fatal illness. Cabot, after a discussion with the farmer's attending physician, a review of the laboratory evidence, and a careful physical examination pronounced him to be suffering from cancer of the stomach. Nine months after this fateful meeting, the farmer wrote back. He noted happily that he was still alive and that he rejected this diagnosis. Cabot had also prescribed some medicine for him to aid digestion in his impaired stomach. Now, nine months later, the farmer had stopped the medicine and felt as good as ever, although he acknowledged some continued pain and reported a loss of weight. Nonetheless, he informed Cabot in his letter, "you made a bad guess" about his condition. In fact, the farmer continued, his own experiments with other medicines had led him to conclude, "I have proved beyond all doubt that I had no cancer. The trouble was the liver. "[27] He too claimed to have followed a careful, reasoned pursuit of information about his condition. But he had reached different conclusions.

He eluded Cabot's disheartening diagnosis of stomach cancer, substituting a diagnosis more to his own liking. Like the patient with catarrh, he also cited support for his conclusions in the process of diagnosis itself, reminding Cabot: "You said that you could not tell the nature of the trouble it was all guess work, to know you would have to look inside which I would not allow & you did not advocate doing so." The former farmer seemed to understand the implications of Cabot's position on diagnosis pretty well. As in the Clinicopathological Conferences, the diagnosis might be the doctor's best opinion based on the available evidence, even when evidence from an autopsy or surgery might be more conclusive. If Cabot could stake a claim to a diagnosis on limited evidence, perhaps the patient could as well. In fairness, as the farmer argued in closing his letter, "If you had wanted to know any thing about my past life, my habits &c. I could tell you more in one minute than [the attending doctor] ever knew about them."[28] Certainly, he was the ultimate authority on the details of his illness.

This high level of involvement in the diagnostic process by these two patients found little representation in Cabot's accompanying notes in their charts. For his office notes, Cabot borrowed the well-established format used in the Clinicopathological Conferences and in the medical charts at Massachusetts General Hospital and similar academically affiliated hospitals of the day.[29] This standard chart represented a process of diagnosis that redefined the patient's difficulties as evidence of disease, making a sharp distinction between the patient's concerns or complaints and the medically interpretable data. In September 1903, for example, a sixty-year-old man, an overseer in an industrial mill near Boston, visited Cabot in his office. This man's experiences, as they were set down in longhand on a fresh page in the record book for 1903, present the evidence of a characteristic diagnostic pain. Cabot had jotted down that the "pain starts in sternum on any exertion—then bores thro' to interscapular region + then into both arms—especially left, then into neck + lower jaw (never upper) . . . never comes on at night . . . makes him keep very still." Information from the physical examination and laboratory testing of blood and urine appeared next. At the bottom of the page, Cabot entered the diagnosis, "angina," based largely on this man's account of his symptoms. Cabot's description of the overseer's pains, in fact, matched neatly with the description of angina in a standard medical textbook of the time.[30] He had translated the mill overseer's story into a medically recognizable account of a disease, angina pectoris.

Any diagnostic efforts that the overseer had contributed beyond the recounting of his symptoms were omitted from Cabot's account. On their face, these charts portrayed a clinical authority that was validated in the dramatic setting of the Clinicopathological Conferences. People provided the messy accounts of their troubles, and diagnostic technique allowed Cabot to redefine them as the evidence of a particular disease. Thus Cabot's term "interscapular region" made the place that the mill overseer had presumably indicated between his shoulder blades into an anatomic region of general medical significance. The chart described a medically relevant pain that began "on any exertion" and not a man who exerted himself in particular ways and felt pain, as he might have characterized it himself. It was the pain that moved, taking its own anatomically defined path from sternum to arms and jaw. This pain, properly understood, was the evidence of a disease, extracted from a story of particular pains and discomforts by the exercise of diagnostic acumen.

This standard medical chart portrayed the process of diagnosis as an internal epistemology for medical work, powerfully legitimating to the professional eye but largely invisible to patients. The rhetoric of these charts, for example, emphasized the distance between patients' descriptions of their condition and the understanding of disease created through Cabot's diagnostic scrutiny. When Cabot directly noted elements of a patient's speech, it was typically in the form of a colorful colloquial phrase, sometimes in quotation marks. The gap between the patient's account and a medically specific translation thereby widened, obscuring any participation in reasoning about disease by the patient. The notes on the mill overseer's angina, for example, conclude with a reminder about the ultimate source of the information. "It's terrible!" Cabot's note says, seeming to quote the overseer about his pain in an evocative but diagnostically inconsequential manner. Another patient, an iron foundry worker, was similarly quoted in the chart to say that his attacks made him so "weak that 'a child of six could knock him out.' " The same man was reported, again in quotes, on a subsequent visit to say that his medicine "relieves him as a dream." The chart seems to highlight the fleeting value of this patient's metaphors—and of his medication—with the next entry dated three months later reading simply, "Dropped dead absolutely sudden."[31]

Cabot's charts presented a diagnostic process in which physicians were distanced from the immediate concerns of their patients by the obligations of attending to disease, and patients were denied participation in the theorizing of their physicians. Patients felt pains for which they sought relief. But instead of responding to the pain directly, the physician interrogated it for clues to the underlying disease. To the diagnosing physician, a patient was not a collaborator in the search for disease but first and foremost a source of naïve information about suffering. The patient described pains and discomforts without attaching any explanation or meaning to them. The medical chart typically reported what the patient felt, rather than what he understood his troubles to mean.

The standard medical chart, perhaps not surprisingly, did an inadequate job of capturing the full dimensions of the exchanges between physician and patient involved in diagnosis. In fact, patients communicated to Cabot a more complex version of their troubles than went into the medical record. Their letters give access to the larger discourses that occurred in this clinic. Patients often interpreted their symptoms not as suffering but as the work of a disease, at least for purposes of communicating with their physician. In their letters, they demonstrate that they recognized and could emulate the process of diagnostic reasoning, struggling to explain their troubles as signs rather than merely as suffering. They indicate sites of internal discomfort, for example, as if directing their doctor's attention to a potential internal site of disease. Undiagnosed patients wrote describing "trouble with bladder" or "something trying to expel itself from my uterus." Borrowing an anatomical, mechanistic model, they might describe their diseases in ordinary language, writing about "colds in chest and accumulation which gathers on the tubes"; or they borrowed more precise medical vocabulary to make their case, such as the young law student who wrote to Cabot that his colds were "largely nasal . . . [with] some inclination to descend to my larynx and bronchial tubes."[32]

Some correspondents attempted their own interpretations of the diagnostic evidence, as if hastening Cabot in his task. Following her husband's visit to the Marlborough Street office, one woman wrote back asking of his condition, "[W]ouldn't he be apt to lose flesh by this time if it was a cancer?" She was unsure what diagnosis had been reached for his condition and had been turning over the evidence herself.[33] The medical chart typically recorded the contributions of patients and their family members only as a kind of naïve sensory data. In their letters, however, people couched their concerns and distress in a language of diseases and disease mechanisms. They were clear about the relevance of their symptoms to the problem of diagnosis. Suffering might be valuable as evidence of a hidden disease. Patients kept this use in mind when they wrote recounting symptoms. A young man, for example, sent a letter in 1909 listing several difficulties that he was experiencing: "Sources in left chest causing pain by use of muscles of that side. Recently there has been little cessation of stinging. Usually increased by eating. Much gas in stomach. Pressure on stomach causes throbbing." He situated pains anatomically and in relation to specific bodily functions. His intention was diagnostic. He had already had Cabot's advice about his condition, and he concluded the letter, signing, "Hoping that these things were anticipated when you diagnosed my case, I am very respectfully. . . ."[34] Did the clues fit the original diagnosis? Loss of appetite and declining weight, or throbbing of the stomach were not just unpleasant; they were diagnostic information, important to patient and physician alike. Like the men suffering from catarrh and liver disease, these patients demonstrated a familiarity with a general model of diagnosis, which Cabot's charts had made to seem opaque to them. Many of the people who came to this office wished to take an active role in the search for their hidden diseases.

The Value of Diagnostic Services

Alfred Worcester warned that the modern disease-oriented physician might wind up supplying a patient with only a diagnosis, rather than with broadly humane attention and moral guidance. Yet for Cabot's patients in private practice, the diagnosis turned out to be an attractive service in itself. Some of the patients who came to the office on Marlborough Street sought explicitly to have their diagnosis first and foremost. There were, of course, people who described no more than a desire to obtain relief from their illness and expressed little concern about the basis of Cabot's recommendations. A few, however, seemed almost more interested in the disease than in his recommendations, in the reasoning behind their doctor's advice than in the advice itself. These diagnostically inclined patients expressed a keen interest in naming their diseases and sometimes also in the evidence available for such naming. Physicians in the early twentieth century were building an authority that derived from and supported their ability to identify, manipulate, and control specific human diseases. Diagnosis provided an epistemological basis for this authority—as was neatly demonstrated in the Clinicopathological Conferences, for example. Yet diagnosis could also function as a discrete, purchasable service, available to individual patients from their private practitioner. Patients did not all come to Cabot expecting to receive a single, undifferentiated package of medical control over their health. The ability to create knowledge about disease through the technical methods of diagnosis was in itself visible to patients and was appealing to a limited extent as a service that they could obtain.

Cabot's practice in fact distinguished itself in specific ways that drew in people who sought the specific service of diagnosis. This attraction was filtered principally through his local professional connections, although not limited by them. Cabot cultivated a name among his peers for his expertise in the use of diagnostic tools. In the 1890s, before his private practice was fully established, he got a start by offering to do analysis of blood samples for other physicians, as described in chapter 2 above. Cabot's growing reputation in the first decades of the twentieth century as an author of textbooks on *Physical Diagnosis* and *Differential Diagnosis* likely also contributed to a growing professional recognition as a skilled diagnostician. Over the first decade of his practice, patients increasingly reached his private office on referral from other physicians.[35] His reputation as a diagnostician may have influenced referring physicians to send patients especially with such questions. When referring doctors wrote to him, they occasionally mentioned the diagnosis as the reason for directing a patient to him.[36]

The diagnosis seemed on some occasions to be exactly what was sought and obtained. Shortly after his wife's visit, for example, one husband wrote back to thank Cabot because "you certainly diagnosed her trouble with absolute accuracy." The diagnostic process might be rapid and persuasive, as in the similarly satisfactory case of a patient who wrote back after being advised that he had a stomach ulcer, to let Cabot know that "for a 'snap' diagnosis you certainly hit me all right and you certainly have my thanks." Other problems were more elusive, but equally illustrated the perception of diagnosis as a valuable service, as suggested by the brief note on a man whom Cabot reported years after a first visit was "still hunting a diagnosis." One patient arrived in Cabot's office in 1899 from Vermont bearing simply a "question of syphilis." After noting the results of a careful interview and examination Cabot entered the diagnosis that this man had traveled so far to find, "syphilis," at the bottom of the chart. Another man, who "thinks he caught syphilis," was told by Cabot that he did not have the disease. This man wrote back several years later politely avoiding mention of the condition, but nonetheless thanking Cabot, since it seemed true "as you stated at that time that you thought it was not what I thought was the trouble." People who went to Cabot might seem quite sure of the nature of the diagnostic questions facing them, like the woman of whom Cabot reported "came for lump in her breast. Much worry."[37] Such patients expressed a belief that the identification of a hidden disease, or of its absence, was a critical service in itself.

These fears about disease could, of course, be tragically well founded. The woman with the concern about the lump did prove to have breast cancer, as she doubtless had feared. The diagnosis was established shortly after her visit through a biopsy of the lump that Cabot arranged with a surgical colleague who had an office nearby in the Back Bay. She went back to the same surgeon for removal of her breast; and after the surgery saw Cabot for postsurgical care, including draining of the surgical wound. Soon she was well enough to go back actively to her work across the river in Cambridge for a short time, but she then began to suffer from what proved to be the extension of the cancer. Within the year, she succumbed to the disease and died at home, with Cabot still involved in her care less than two years after her first worried visit. The case concluded in Cabot's record books with the evidence cited from an autopsy on her body demonstrating the extensive spread of cancerous tissue through her chest and spine: a sad and eerie confirmation through the diagnostic enterprise of her initial concerns about the lump.[38]

Many patients seemed to acknowledge the importance of a diagnosis. A diagnosis after all could serve many purposes, even beyond confirming or dispelling a fear about disease. Diagnosis was, for example, the key that could unlock access to desired medical treatment, as patients themselves sometimes represented it. The assumption that the diagnosis necessarily guided treatment was fundamental to physicians like Cabot who hoped to adhere to a medical dogma of discrete diseases with specific, well-defined therapies. This dogma might seem to hold little value to someone who simply sought relief from an illness. Yet patients too accepted the premise that a diagnosis of disease determined treatment sometimes as readily as their physicians, and attempted to apply this logic to their own care on occasion. A woman from the nearby city of Dorchester wrote to Cabot in 1913 recalling an earlier visit to his office. She had come to him with the concern that she had syphilis and "to ask you to help me to be treated with the new remedy 606." She understood quite well that the new medicine 606, or Salvarsan, was the magic bullet against this dread disease in 1913. She also recognized that it was the diagnosed condition of syphilis that would qualify her for treatment with 606. She reminded Cabot how he had withheld 606 from her, since "you said that you did not think that I had the disease." She had gone subsequently, she reported, to another physician, who also told her that she did not need the medicine because she did not have syphilis. In this instance, she recalled, the physician had sent her to get a Wassermann serum test for syphilis, "to convince me" about the absence of the disease.[39] She disagreed with her doctors about her diagnosis and proper care. But she accepted, at least for the sake of their discussion, the doctors' premise that the diagnosis of syphilis was a requirement for treatment with 606. Thus she seemed not to pursue the argument that 606 might be good for her in any case and instead criticized the physician's use of a Wassermann test "to convince," rather than to diagnose accurately. Her argument turned on the question of the presence of syphilis and not on her right of access to the medication she wanted.

Patients vividly represented other purposes for the diagnosis in their correspondence as well. Some people expressed their troubles to their doctor in shorthand, according to the name of a disease. They referred in their letters to a great variety of conditions that doctors, and others, had named for them, like "anemia," "malaria," "gastritis," "eczema," "tonsillitis," "renal calculus," or "Addison's disease." Some patients conjured up nothing more specific than a difficulty with "spells." Or they described their condition in purely symptomatic terms—for example, "curious attacks of chills and exhaustion." But in general people strove to identify their conditions with diagnostic terminology, both conventional and otherwise, and to interpret their troubles for their physicians as the manifestations of disease. It was surprisingly rare for Cabot's patients to suggest in their letters that they were afflicted with conditions that had no identifiable relevance to medical knowledge. Patients who described their conditions as "catarrh" or "Grippe" used an older medical terminology, but one that was easily translated directly into the updated terms that were in use among Cabot and his peers. Purely slang terms for illnesses were, of course, in widespread use elsewhere, but they made little appearance in these patients' letters, targeted as they were at a specific medical audience, the doctor.[40]

An understanding of diagnosis might seem especially crucial in negotiating with more than one physician. A diagnosis was recognizably the key to understanding medical justifications and rationale, and it likewise served to communicate information between different physicians. Physicians like Cabot might elicit this need in a patient, as in the instance of a young woman who wrote after a visit in 1904 that "it is very difficult when a patient does not understand technical terms and until today I never would get Dr. Perry to talk it frankly out with me." Knowing that a diagnosis had a specific relevance for her was brought out in her meeting with Cabot. On the occasion of the visit, Cabot asked the reason for a previous surgery that she had undergone at Massachusetts General Hospital. He noted in her chart that she "doesn't know for what hysterectomy and appendectomy." The "what" would presumably be the diagnosis, the disease that removal of her appendix and uterus had been intended to treat.[41]

A diagnosis was a convenient means of referring not only to a specific rationale for care but to an entire package of medical services. Patients found an advantage, for example, in the ability to characterize their therapy directly in terms of the disease that they harbored. The patient mentioned above who knew that he had anemia recounted how he had received treatment from another physician for that condition; the woman with eczema expressed appreciation for Cabot's referral to a helpful specialist in skin diseases for this condition; and the woman who described stomach troubles wrote that she had recently found a new doctor who "rid me of gastritis for a year and a half." The man with a self-reported "renal calculus," or kidney stone, assured Cabot that he had solved this problem himself with a long bicycle ride, which he claimed had knocked the stone loose through "a good shaking up."[42]

Not everyone made a tight connection between diagnosis and treatment. Many of the people who found their way to Cabot's office accepted their doctor's interpretation of their problems tacitly as an indistinguishable part of a more general responsibility for medical care. They may have assumed quite reasonably that a doctor who presumed to treat them knew what was wrong—more often than they stated explicitly in their letters. So one correspondent asked Cabot merely to "send me news of a 'serum' that would fix this trouble," without inquiring more about what Cabot actually thought that the trouble was.[43] The general request to fix things up was not necessarily prefaced by a discussion of the nature of the disorders that were being fixed.

The strongly expressed interests in specific diseases among Cabot's corresponding patients did emerge from a background of more elaborate and contingent observations on personal health. Patients wove their mention of disease into longer narratives, suggesting other underlying conceptions of ill health as a more general condition inseparable from the details of one's life, habits, history, and surroundings.[44] People addressing themselves to Cabot offered up a great variety of observations for use in sorting out their troubles. "I can go [on] walks—play tennis do all my own work but *washing*, but just a little too much of anything especially church work seems to undo me," reported one correspondent. General observations assuring Cabot, for example, that "my condition is fairly good at the present time," or that "the walk to your office evidently did me good for I had a real appetite for dinner," alternated with concerns about particular environments and activities, as "if I eat sweets or apples I am sure to pay for it" or "cold weather does not agree with me."[45] Discrete disease was only one possible culprit in ill health, connected to general notions about the many potential influences on individual well-being. Disease was one influence, perhaps not surprisingly, that patients seemed particularly eager to discuss with their doctor.

Diagnosis and the Limits of Medical Authority

In his essay on "The Care of the Patient," Francis Peabody cautioned doctors against concluding prematurely that there was no disease in a patient who tested out normally. Peabody advised physicians not to constrain their attention and care too stingily according to the evidence of specific diseases. People without a diagnosis needed support and monitoring as well. Richard Cabot seemed to harbor the opposite concern: that undiagnosed or vaguely defined illnesses might serve as an illegitimate pretext for a physician's ongoing medical control and intervention. He used the requirement to base medical attention on a clearly defined diagnosis as an important restraint on the doctor's authority. A well-founded diagnosis created the basis for a physician's recommendations and interventions. Equally, the requirement that physicians would actually establish a diagnosis first might bar them from extending their influence into areas where it had no basis and might be detrimental. Cabot's was not a typical concern of his day, but an illuminating observation on the emerging shape of twentieth-century diagnostic responsibilities.

The diagnosis of the absence of disease thus had serious implications for Cabot in his ideal vision of medical practice. This diagnosis of "not ill," as he used it in his office clinic, marked the boundaries of valid medical control, especially in areas where it could be contested. For example, Cabot argued that under the private, fee-based medical system in which he and his colleagues practiced, the discovery of disease equated closely with the recruitment of new business. Cabot speculated publicly, and controversially, that doctors were biased against finding too many of their potential customers well. As he put it provocatively in a muckraking article in the *American Magazine* in 1916, "some patients do not like to be told that nothing is the matter with them" and the doctor dependent on them for income was unlikely to disappoint.[46] By linking medical care tightly to the identification of objectively defined disease, the medical profession could restrain excessive promotion of medical services by individual doctors, he suggested.

Cabot was favorably impressed by the notorious denunciation of the perverse incentives in commercial medical practice by his contemporary George Bernard Shaw. He found a satisfying aptness in Shaw's musings about the well-deserved fate of anyone who gave a surgeon "a pecuniary interest in cutting off one's leg."[47] The greatest satisfaction a doctor could have, in Cabot's atypical expression, was the pleasure in telling people that he was not needed. He suspected, no doubt rightly, that few of his medical colleagues shared this pleasure, or at least felt that they could afford to. He did attract spirited support from people outside of medicine, both in his time and subsequently, who also saw in his critique evidence for a crucial problem in commercial medical practice. Cabot documented bluntly in his office charts the cases in which he believed that a patient's ills had been "increased and largely created by [another] doctor," typically through the suggestion that diseases existed where he felt that they did not.[48] His personal experience of patients eager for diagnostic attention may have encouraged him to consider this a significant problem. Public enthusiasm for the power to diagnose seemed to him more difficult to restrain than to recruit.

Cabot as usual attempted to practice what he preached by extending the diagnosis of "not ill" to a small, slightly increasing fraction of the people who came to his private office, especially after about 1910. His careful practice records showed how this diagnosis functioned and how some of his patients responded to it in their correspondence. People who received the diagnosis of no disease generally did not return for future examinations and consultation.[49] Cabot did not follow the strategy that Francis Peabody would recommend. True to his principles, he seems to have excused people whom he diagnosed as not sick from the need for his further care or attention, although in the case of patients who were referred to him, the unstated assumption may have been that they would return to their original physician. Cabot's strategy was not always easy to put into effect, since most people who came to the office did so because they believed that they were sick. Routine examinations for healthy people were not a part of his practice or widely in use until much later in American medicine, so it was rare for someone to come feeling well and expecting to be proven free of disease.[50] Nonetheless, Cabot diagnosed no disease in almost one of every fourteen new cases at a peak rate in 1915 (see table 5.1 below.)

The diagnosis of no disease for Cabot meant that the patient's troubles not only lay outside his responsibility but might lie outside of the direct responsibility and control of any physician. Absence of disease marked a potential boundary where the physician might pass off expert care to others. Cabot's previously described reforms in medical social work and religious counseling figured neatly into this strategy. His controversial work with the Emmanuel Movement between 1906 and 1909 is a good example. The charismatic leader of this religious healing movement, Rev. Elwood Worcester, promoted spiritual counseling for people afflicted with nervous disorders, using nonphysicians to conduct group meetings with prayer, autosuggestion, and hypnotism. Cabot initially collaborated with the project, certifying that the patients cared for by Worcester's group had no specific medical diseases causing their difficulties. Worcester had eagerly endorsed this function of modern diagnostic acumen, which seemed to him to define a group of people well suited to his purpose. In his foundational book on the movement, Worcester explained that an "important characteristic of our work is the pains taken in the diagnosis of disease." He argued that a careful diagnostic evaluation of everyone entering his program would protect both the patients and the spiritual practitioners, since "so-called nervous affections are not infrequently indications or precursors of serious organic disease, failure to discover which . . . is frequently a gross wrong to the patient and the exposure of oneself to merited criticism and contempt."[51] Although Worcester's system made sense in the context of Cabot's practice, it exposed him to the criticism and contempt of the medical profession at large. As it gained national popularity, the Emmanuel Movement increasingly drew the ire of physicians concerned about the loss of their patients to what one angry practitioner characterized as "Emmanuel Movement, new thought, Christian Science, magnetic healing, in fact the whole shameless rabble."[52] Cabot himself backed out of his involvement with Worcester and the group in its waning days in early 1909.

Patients who came to Cabot's clinic quickly registered the idea that the diagnosis of no disease meant that there might equally be no specific obligation for the physician. Their expressed reactions to the diagnosis differed sometimes strikingly. Sometimes it had a reassuring influence. One woman wrote to thank Cabot for telling her that she was free of threatening disease. She attributed her recovery gratefully to "the day when I insisted upon your telling me whether or not there was a growth in my stomach." For others, the definitive opinion that there was no disease seemed to imply abandonment. Another woman explained poignantly to Cabot that she understood him to mean that she was being sent away to "work out my own salvation"; perhaps, she allowed, Cabot "had hoped you might benefit me." The diagnosis of no disease might also signify the refusal of sought-after medical care. It was the diagnosis that the woman who came seeking 606 from Cabot had received when he determined that she did not have syphilis.[53] She had gone elsewhere, she informed Cabot in her subsequent letter, maybe seeking better agreement about treatment or perhaps just the Salvarsan.

Cabot seemed to want the diagnosis of no disease to serve as a boundary marking the limits of valid medical influence and intervention. Yet the patients in his office brought their own interpretations and interests to the work of diagnosis. In 1914, a middle-aged man arrived from Vermont bearing a story of several years of troubles including, Cabot noted, "much pain in belly with no relation to food." Concerned about the possibility of disease but unwilling to diagnose the case without more information, Cabot sent this man over to Massachusetts General Hospital for further evaluation and testing. His stay at the hospital lasted a week, during which time the Vermonter underwent a variety of blood tests, urine tests, a serum test for syphilis, stool tests, and serial chemical measurements of kidney functioning. He was persuaded to swallow a tube so that fluid could be sampled from his stomach for chemical analysis before and after he ate a standard test meal. He drank a dose of liquid bismuth and had x-ray photographs successively documenting the transit of the bismuth through his stomach and bowels. The hospital physicians, writing to Cabot in summary, noted that their investigations had turned up nothing conclusive. They had, in fact, attempted to obtain more diagnostic information. They proposed using an enema of barium followed by more x-rays. The patient, however, proved to be "in a hurry to get away and did not stay," they explained, after a week of such attention.[54] The doctors offered no final diagnosis in their report.

Writing back to Cabot soon after her husband's return, the Vermont man's wife confided to Cabot that ever since his hospital experience, "he is quite sure that you made up your mind it was a cancer." A lengthier search perhaps implied a worse diagnosis in this patient's mind. The wife told Cabot that she herself was anxious and could not "wait any longer without knowing something more definite." The doctors seemed unsure of whether they had looked hard enough for disease, but were unable to convince the patient to look harder. The patient was reportedly sure only of what it was that he thought his doctors were looking for and hoped for "something more definite."[55] The only matter on which all the participants were firmly agreed was their concern about the responsibility for a hidden disease.

THE REMARKABLE GROWTH of the physician's persuasive authority in the early twentieth century has yet to find a full historical account.[56] For Cabot and some of his colleagues, the increasingly complex processes of diagnosis posed important questions about the definition and exercise of this authority. Alfred Worcester and Francis Peabody expressed concern about the influence that the diagnostic process itself exerted on their patients. Both argued that doctors had to learn to moderate and guide diagnosis carefully to avoid detrimental effects. Perhaps the diagnosis of a disease should be avoided in some cases, as Worcester argued. Or, as Peabody suggested, it could be postponed indefinitely, and the process of diagnosis drawn out into a form of benevolent vigilance. Diagnostic quandaries were not new to medicine, nor were concerns about the nature of individual medical influence and control. Peabody, however, seemed to find a novelty in the expectation that a week of sequestered testing in a hospital might result in no clear understanding of what was wrong. He argued that such events called for special strategies in managing the process of diagnosis.

Cabot approached similar concerns with a different conception of the influence properly exerted in the work of diagnosis. The diagnostic process could, in an ideal formulation, lift the doctor above the potentially confusing accounts and concerns of the individual sick person to an exacting consideration of the evidence of disease. Cabot contrived a compelling and enduring demonstration of this power in his Clinicopathological Conferences, where the patient was absent and the physician was tested against a professional standard of the autopsy. Cabot relied on his convictions about the validity of this process to support a policy of telling some people who felt quite sick that they were in fact "not ill." Adherence to the rational pursuit of disease provided a secure basis for the doctor's independent authority and a potential limitation on the illegitimate extension of this authority into the management of patients outside of the care of specific disease.

In his private office, away from the sequestered realm of the Clinicopathological Conferences, Cabot's authority drew more direct support and challenge from his patients. The people who came to him found ways to introduce their own interpretations and needs into the diagnostic process. Diagnosis, far from remaining an implicit element of the doctor's independent pursuit of diseases, seemed a visible, sometimes distinctly valuable service. The hunt for hidden diseases proved attractive in itself, especially if it provided the right result. The man from Vermont endured a week of testing apparently convinced that there was some problem that his doctors could at least identify for him. People who went to Cabot might willingly endorse the value of diagnostic services. But they could also express more interest in the diagnosis than in the physician's general advice and care. Some patients even demonstrated an enthusiasm for the diagnostic process that outstripped their willingness to yield to the control of the physician who was ostensibly directing it.

CHAPTER 4

TREATMENT

How to Know What Works

W hat kinds of agreement could early twentieth-century physicians reach with their patients concerning treatment? And what inducements could they provide for patients to accept increasingly potent and risky therapies? Doctors long before Cabot had been accustomed to seek trust and support for their therapeutic plans in many sources, including local reputation, agreement on therapeutic mechanisms, established relationships with patients and families, and sustained attention over the course of an illness. But risky and potent treatment like aseptic surgery or chemotherapeutics more routinely raised the stakes in therapy. The patient who came to Cabot seeking the new treatment called Salvarsan, or 606 (chapter 3), was frustrated by her inability to find anyone to administer it to her. Cabot tried to explain that she did not need the medication, since she did not test positive for syphilis. But she continued her search. She went to another physician, who wrote to Cabot in exasperation after his first meeting with her, noting that "somehow the desire for 606 has fastened itself upon her and I fear that she will not be happy till she gets it."[1] Salvarsan was moderately toxic, and he feared that she might have the misfortune to find another physician who would indeed prescribe it. This patient's struggle over access to the medication pointed to certain anxieties about therapeutics that were becoming more familiar to Cabot and his colleagues in the early twentieth century. Patients might eagerly seek special new treatments, but what basis did they have for judging their appropriateness or their effects? Since medicines like 606 were potentially dangerous, the doctor's influence over their use might entail heavier responsibilities and more stringent criteria for treatment.

Physicians increasingly sought to confirm the effects of therapy through specialized techniques of physiological monitoring, making the patient's immediate experience of treatment seem more incidental. As doctors adopted powerful new disease-modifying treatments, they took to gently criticizing traditional treatments that they felt were "merely symptomatic"— as the writer on "Therapeutics" in the 11th edition of the *Encyclopedia Britannica* put it. How would physicians conscript patients into treatment that was assessed and adjusted, not according to symptoms, but according to hidden markers for disease? Such physiologically guided therapeutics seemed to demand new reassurances from the treating doctor. One hope among the innovative physicians like Cabot who urged these changes was that the demonstrable power of these treatments would bolster the authority to guide them. Sustained cures would mean happy and compliant patients. In many cases, however, these medications made only transient or partial improvements in how the patient felt. Cabot and many of his peers began to offer the evidence of therapeutic monitoring itself as a reassurance and confirmation about the value of treatment. Testing to monitor and assess the course of a disease under treatment became one means of demonstrating to the patient the effectiveness of medications or procedures whose therapeutic benefits might otherwise seem ambiguous or slow to arrive.

Among Cabot's patients, a parallel set of concerns was emerging. What did it mean to get better from a treatment? With growing public information about miraculous treatments like diphtheria antitoxin and 606, how did a person know which treatment was the right one? There were so many new therapies and therapists to choose from in a growing health-care marketplace. What constituted adequate evidence that a treatment indeed worked? Just feeling better seemed at times inadequate proof of effective treatment, just as feeling worse was judged compatible, at least at first, with therapeutic success. As one young man wrote to Cabot describing the outcome of his surgery, "the operation itself was a success," he reported, "but I have never recovered from the check to my nervous system."[2] He identified a significant difference between the medical success of the "operation itself" and his mixed experience of treatment and recovery. Rather than discounting his acceptance of a medical definition of success as simple pandering, we might inquire further how he, and others, used medical definitions of success—even as they registered the incompleteness of these medical answers to the challenges of getting better.

Symptomatic and Physiologic Therapy

Therapeutics in the late nineteenth and early twentieth century began to seek a strange forbearance from patients. It was not that medical treatments were becoming more unpleasant or noxious, although Cabot's patients did describe some difficult experiences of care. Nineteenth-century physicians had similarly asked patients to endure some very taxing treatments on occasion, especially when the medical condition was itself severe. But a gap opened toward the end of the nineteenth century between a patient's experience of medical treatment and the treatment's intended effects. The targets sought out by newer twentieth-century therapeutics often had little connection to their immediately perceptible effects. Cabot wrote of a patient with pernicious anemia in 1912 (chapter 1) that "when his red [blood] cells begin to fall, that fall can be retarded by giving Fowler's Solution beginning with two drops after each meal and increasing up to the limit of toleration, that is until nausea and diarrhea . . . are produced."[3] Fowler's Solution aimed to retard the fall in the count of red blood cells and only incidentally had the effect of making this man feel nauseous. The ultimate goal, of course, was to treat his illness, the pernicious anemia. Yet Cabot and the physician to whom he wrote were willing to relegate the patient's immediate experience of the treatment to the role of an unpleasant but medically incidental nuisance. The perceptible effect of nausea was unrelated to the drug's true therapeutic action. In contrast, a delay in the predicted fall in the count of red blood cells was almost by definition undetectable for the patient. Patients and doctors earlier in the nineteenth century had indeed dealt with some troublesome effects from therapies. Quinine, a mainstay of nineteenth-century therapeutics among Cabot's predecessors, was known to cause a disturbing ringing in the ears in doses large enough to suppress fever, for example.[4] The twentieth century, however, marked the deployment of a therapeutic rationale in which perceptible effects were sharply distinguished from therapeutic actions, and where therapeutic effects might only be evident through specialized physiological monitoring.

Medical therapy had chosen targets deep within the body that were suddenly made visible by technical means. The idea of guiding treatment according to the medical understanding of bodily function had ancient roots in medical theory. What changed late in the nineteenth century was the availability of routine methods for extracting information about physiological effects: with blood counts and hemoglobinometry, chemical urinalysis, microscopy, serology, and x-rays. One attraction of physiological monitoring may have been a kind of hermeneutics that it provided—by revealing in laboratory results the otherwise hidden significance of a medical treatment. But physiological therapeutics also had a compelling simplicity to their rationale. Count the number of red blood cells and determine if it rose or fell with treatment. Any subtleties or difficulties with this effort could be attributed to the monitoring practice: gathering an adequate sample of blood, identifying red cells, and partitioning and counting them. The application of the information to treatment then seemed obvious, at least in general outline. It was a rationale that appealed not only to Cabot and his peers but to some of their patients as well.

Many of Cabot's treatments were medicinal; he gave pills and tonics. But his therapeutics was likely influenced by the success of his surgical colleagues. In surgery, the commonsense justifications for therapy often had to overwhelm a patient's immediate cautions and speculations about the nature of therapeutic effects. Removing the diseased part made sense at some basic level, despite the accompanying terrors of cutting and sewing. Abdominal surgery was a particularly decisive case, in which the internal defect was both made visible and corrected through a single dramatic procedure. Chapter 1 noted how, in 1910, Hugh Cabot showed a mother her son's appendix, which he had just excised in an adjacent room, on a napkin.[5] Physicians might need only share such surgical evidences rarely before developing a sturdy confidence in their ability to gain approval for treatment. This confidence in the evidence of therapeutic assessment and monitoring was neatly evident in the correspondence among Cabot's peers relating to both medicinal treatments and surgery.

Drug therapy increasingly sought to produce interior changes in a patient's body that were as concrete as surgical effects. Dr. Sarah Bond in 1915 summed up to Cabot her treatment of a patient's anemia in a brief letter. This patient's hemoglobin, she indicated, measured "about 50% and it has not improved under iron and arsenic which she [her patient] has been taking for several weeks."[6] Dr. Bond communicated nothing more to Cabot about the state of her patient beyond the measurements of the hemoglobin, a constituent of the blood visualized with a tool called a hemoglobinometer. She was not alone in placing such weight on a simple technical measure of therapeutic progress. Another physician wrote to Cabot saying of his patient that "his symptoms have all been objective" and describing those symptoms—the disordered microscopic and chemical characteristics of the man's urine—in detail. This doctor went on to describe various medical treatments that he had tried, concluding with the report that the treatments had "acted favorably on the renal lesion," as evidenced by the improving microscopic quality of the urine.[7] Testing and assessment of red cells, hemoglobin, or urinary constituents provided a persuasive material reality for the internal targets of therapy, at least among physicians. This form of professional agreement about markers of therapeutic success was a distinct change from the concerns of mid-nineteenth-century medicine.

Attention to the criteria of physiological therapy displaced the nineteenth-century reliance on the experiences of patients in assessing treatment. The change was evident to physicians who lived through it. Writing about physiological therapeutics in 1902, the prominent American physician Nathan Davis recalled how in the nineteenth century, "the effects of remedial agents were determined by their visible influences on the various evacuations and on the sensations of the patient."[8] He contrasted this older mode of treatment with more modern efforts to monitor and assess treatments through their measured physiologic effects. Nineteenth-century therapy with mercury, belladonna, skin plasters, alcohol, morphine, lobelia, or bloodletting was intended to alter the pulse, trigger purging or vomiting, calm the nerves and pain, draw blood to the skin, dry up or augment the saliva, or increase urination. These obvious changes in "the various evacuations and . . . sensations of the patient" were the sought-after healing effects of the treatments. The ability to adjust medications to create these desired bodily effects in a patient was the essence of treatment.[9]

Therapeutic Monitoring and the Patient-Doctor Relationship

Physiological therapeutics established, and required, a greater asymmetry between the twentieth-century therapist and patient. Nineteenth-century treatment gave the patient greater control over therapeutic evidence, because a doctor had to seek any confirmation in the report of sensations and bodily responses. Feeling better was never the sole prescribed route to getting better, of course; but how a patient felt still mattered first and foremost. Patients in the nineteenth century sometimes endured long courses of bloodletting and dosing with mercury that were fully as vexing as anything recommended by the twentieth-century physician. Nor were nineteenth-century therapeutics lacking in the theoretical complexities that obscured the rationale of the physician's craft. Nineteenth-century treatment relied on theoretical knowledge about disease processes and bodily reactions wholly as subtle and variegated as anything cited in physiological therapeutics. Control over the levels and flux of urine, saliva, blood, bowel movements, or bile through the body aimed to reestablish the obscure internal balance of flows and pressures that was upset by different diseases. Nineteenth-century physicians claimed a special expertise in recognizing and correcting these imbalances. Yet the arduous path of nineteenth-century treatment remained more overtly intelligible and perceptible to the patient who traveled it.

By contrast, new criteria for physiological testing posed obvious challenges for the prescribing physician. Early twentieth-century supporters of physiological therapy expressed concern about the implications for their therapeutic relationships with patients. Powerful new medicines that struck effectively at disease might draw patients to the physician's side. But since feeling better was not necessarily or immediately related to real therapeutic effects, the negotiation of ongoing treatment could be problematic. Consider again the treatment of syphilis with the powerful medication 606, or Salvarsan, which Cabot's patient had unsuccessfully sought from him. Charles Whitney, a medical colleague of Cabot's in Boston, speculated in an essay in 1916 that the effectiveness of a drug like 606 for syphilis undermined the doctor's control over treatment, rather than enhancing it.[10] In the early stages of syphilis, the initial dose of 606 sometimes produced rapid, superficial improvements in a patient's syphilitic sores and swellings. These improvements were, however, deceptive. The microscopic syphilitic organisms advanced more quietly after a single, partial treatment, and physicians found that they were "not always able to convince the patient that he needs further treatment and he drifts away feeling assured that he is as well as ever."[11] Before the discovery of Salvarsan, Whitney recalled, a typical patient being treated with established medications like mercury and antimony more readily offered "diligent and faithful cooperation," understanding that "a cure required a long course of treatment faithfully followed." Salvarsan provided falsely reassuring results that undermined the ability to guide patients through a complete treatment.[12] Feeling better was not accurate evidence of therapeutic effect. Whitney cited his own practical experience to argue that such treatments offered no easy leverage to the doctor seeking to influence a patient's acceptance of a course of treatment.

Other physicians of the time spied a solution to this problem within the process of physiological monitoring itself. Assessment and monitoring of disease could become part of the process of securing ongoing cooperation with treatment. Promoting such new therapeutic norms in 1912, George Dock advised that patients had to be "taught that the remedy prescribed is only part of the treatment, [and] that trained intelligence must accompany them until well."[13] Physiological monitoring represented to Dock an ideal means to control the process of therapy through continued "observation and examination under treatment." Since the physician could now, for example, "see the pathologic changes in the blood being treated," it was no longer acceptable, he warned, to send a patient off with a medicine and the simple advice "to return if he does not get better."[14] Trained intelligence, as Dock put it, must accompany them through the process. Control of treatment required control also over the assessment of its effects. A doctor who left the patient to decide independently about the results had failed in an important professional duty, since there was, Dock contended, "no real difference between this and the self-medication based on newspaper advertisements or druggist's posters."[15] A patient could not easily be stopped from purchasing one or another fashionable remedy from the local druggist to see if it worked; but someone who came to the physician for care should expect to be tested as a way of following and confirming the treatments. Seeming to take Dock's lead, Charles Whitney held out a similar solution to the problem of controlling treatments with Salvarsan. The blood test for syphilis, Whitney argued, "is of the greatest value to us in verifying the results of our treatment, and especially in showing the patients . . . that the disease is still present" when they resisted further treatment.[16] A blood test, like an appendix on a napkin, showed both the hidden target of the treatment and its demonstrable therapeutic effects.

Physicians in the early twentieth century were beginning to gain the ability to control a patients' access to medications through their individual prescribing practices. But with few exceptions, any identifiable medicine that was available from the doctor was available elsewhere, as well as a wide array of substitutes. Physicians faced competition from many sides, from alternate purveyors of conventional treatments, from alternate purveyors of unconventional medicines, and from their own inability to enforce the distinctions between them. The question of who got medicine, and how, had practical significance for every physician in private practice. It was not surprising then that the early twentieth-century medical literature tended to characterize patients who came seeking treatment as rather crude consumers in need of careful professional oversight. In essays and editorials about therapeutics, Cabot's colleagues tended to present a consistent image of the typically fickle customer for treatment. Writing about practice, they noted the challenge of a public that "was anxious for a quick, sure, and easy cure."[17] Patients were seen as liable to make shallowly informed choices about their care. They searched indiscriminately for cures, valued treatments only according to their immediate effects, and offered no loyalty to anything that failed these tests. The pressure to treat the immediate symptoms of the disease, Dr. Albert Geyser complained in 1916, came from the patient, who "expects results cheaply and quickly and if it is not forthcoming from one doctor then the patient tries another . . . [since] nothing but visible, immediate results count."[18] In contrast, as Whitney had noted in regard to syphilis, patients who saw quick, superficial results were then wont to drop all treatment, since "seeing nothing they fear nothing and are therefore unwilling to be called sick."[19] They needed to learn that the physician's monitoring of treatment provided the only sure answers about effect.

Patients as Fickle Consumers

The shortcomings of the average patient had long been a theme in medicine's professional orations and essays. Airing these concerns obviously served solidarity among doctors, emphasizing a shared challenge in applying their expertise, rather than their disparate obligations to their many clients.[20] But this critique of the fickle patient held special relevance for Cabot and other physicians concerned about the management of physiological therapy.

A marvelous heterogeneity of therapeutic practices and products greeted the person who was looking for care around New England in the first decades of the twentieth century. Patients who wrote to Cabot recounted their experiences with a great variety of resources and only rarely expressed any frustration in getting specific therapies that they sought. In this respect, the woman who was seeking 606 was a rare exception. Patients described many treatments offered to them by Cabot's conventional medical peers and their competitors. Although they reported on occasion no more than that they had received "Medicine" or a vague "treatment," they also named specific therapies, ranging from "the use of x-ray" to "forty tablets of 'Erythrol tetranitrate' " to "an operation (laparotomy) at the Boston City Hospital."[21] In addition, patients told of finding their way to mind-cure treatment, "an osteopathic treatment every week," homeopathists, an "herb doctor," and Christian Science healers.[22] They also described their own experiments with static electricity, "an electrical instrument called the 'Neurotone,' " Warner's Self-Cure, and yogurt capsules.[23] Many things were available in this market for treatment. Patients faced complicated choices, and those who had the requisite resources could move from treatment to treatment in search of a durable fix. The man who wrote saying that he was taking yogurt capsules also told Cabot that he had gotten a prescription for a medicine from one of his neighbors, presumably another Back Bay practitioner. In addition, he wanted to know if there was "anything new under the sun" that Cabot could prescribe, and requested further whether there was likely to be "any virtue in Goat lymph?"[24] It was difficult to know whether the next treatment offered the key to lasting relief.

Popular demand for treatment, of course, offered a reasonable means of recruiting patients to an office like Cabot's, but only if the patient's quest could be captured and channeled onto conventional paths. According to the professional ideals of the day, physicians would undertake to provide stable, ongoing exchanges and responsibilities in treatment, rather than the simple remittance of a box of pills on request—although this kind of reflex prescribing seems to have been sufficiently common to warrant frequent parody in the medical literature of the day. The literature was as critical of the physician who merely prescribed on request as it was of the patient who merely requested. The notion that treatments like goat lymph, the Neurotone, or erythrol tetranitrate might be available simply for the asking chafed against the values of professional control. Physicians' complaints about their restless consumers asserted a desire to control therapy and a parallel concern about an open market for treatment that could undermine this control. We can see the American medical profession's struggles against this market being played out in the early twentieth-century campaigns against commercial, proprietary medications.[25] Physicians pressed their colleagues and the public to empower them to establish sharp distinctions between Salvarsan and Warner's Self-Cure, or between the mail-order "Neurotone" and the surgical laparotomy, and to control access to the therapies that they endorsed more tightly.

These anxieties played out with special fervor in the debates of the day over physiological therapy, and monitoring seemed at least a partial solution. Albert Geyser, in his warning about the patient "who expects results quickly and cheaply," contrasted the quick relief that patients sought with the "physiologic treatment" that physicians should properly apply. The opposite of physiological treatment in Geyser's account was "symptomatic treatment," the unreflective remediation of symptoms.[26] The establishment of physiological parameters like measured hemoglobin or urinary chemical constituents as proper criteria for therapy left doctors with an uncomfortable question: "[W]hat is a symptom and to what extent is symptomatic treatment rational or permissible?"[27] Treating symptoms might be necessary for a humane practice or to cultivate the cooperation of patients, but the stronger professional endorsement of physiological goals sometimes lent an apologetic tone to this discussion. One physician said that he was "compelled to rely largely upon what for want of a better name we term symptomatic treatment; and up to a certain point and within certain limits this is an essential factor in successful practice."[28] Practice required the ability to provide treatment and the largest part of the traditional pharmacopoeia of the early twentieth century still offered medications like soporifics, anodynes, and stimulants that were known, as the categories themselves implied, primarily for their effects on perceptible bodily function. A reliance on the treatment of symptoms was part of the well-established legacy of nineteenth-century practice, and while this reliance could be displaced, it could not be casually ignored.

What the twentieth-century physician needed in part was a new justification for symptomatic therapeutics. One such defense aimed curiously to free it from the suspicion that it simply appeased demanding patients—the fickle consumers of medical services. Occasionally, doctors defended "the immediate relief conferred" in treating symptoms as simple compassion.[29]

After all, a practically minded editorialist in the journal *American Medicine* noted in 1903, "the patient does not come to us to pay for our theories and diagnoses, but to be made well."[30] Yet such accommodation risked pandering to the unexamined requests of patients. Other physicians pointed to a deeper rationale for symptomatic treatment, emphasizing that the fundamental aim was not merely to palliate disturbing symptoms. Cabot added a minor voice to this chorus, arguing that treating symptoms was warranted, because it was likely to improve a patient's own physiological responses to disease.[31] Physicians need not feel defensive about treating symptoms, since they were actually "supporting, opposing, imitating or altering the natural bodily responses to disease"[32]—although in practice they ended up simply dealing with the most obvious bodily responses, that is, the symptoms. So a cough, Cabot explained, should be suppressed well enough to keep it from interfering with recuperative rest and nutrition, but not so much that it ceased to expel diseased material from the lung. Palliating a symptom was acceptable, but not as an end in itself. A similarly apologetic tone also intruded on the discussion of placebo medications in this period, a debate that involved Cabot as a central figure, as I shall describe later. The literature on placebos similarly portrayed patients as crude consumers challenging the doctor's ability to exert appropriate control over therapeutics.

A change in the contents of the doctor's black bag seemed to call for new norms to guide treatment. Physiological therapeutics picked out interior targets for medical therapy in a manner that made a previous reliance on the patient's perception of therapy seem less legitimate. Hobart Hare in his 1898 textbook on practical therapeutics described the campaign for therapeutic progress in colorful terms, advising that "the old-fashioned 'shot-gun' prescription containing many ingredients . . . should be supplanted by the small-calibre rifle-ball sent with directness at the condition." This metaphor was compelling enough to find wide use, being repeated almost word for word in Wallace Abbot's article "Plea for a Truer Therapy" in 1903.[33] The metaphor served simultaneously in several capacities. New medical treatments seemed to these observers more exact and more potent. No longer would doctors spray mixtures of medications at a problem, judging their results by the obvious changes that they made in a patient's bodily evacuations or sensations. Modern medical treatments, like the rifle bullet, were precisely aimed and deadly, although their targets were not always superficially evident to the patient. What should a doctor expect of patients in targeting obscure internal processes and diseases? Physicians schooled in medicine's ancient literate traditions could quote from their Hippocrates the ideal that "the patient ought to side with the doctor against the disease."[34] In a new twist, however, twentieth-century physicians seemed to be asking that the patient should just hold very still to let the doctor get off a better shot at the disease. Physicians had to consider how to cultivate this kind of cooperation among the people who came to them seeking treatment.

Prognosis as a Means to Cooperation

Many of the existing means for gaining therapeutic influence over patients were unrelated to physiological monitoring, although not necessarily incompatible with it. For a purist like Cabot, however, certain common means for encouraging cooperation with treatment seemed to undermine a reliance on monitoring. He graciously documented common practices of therapeutic persuasion among his colleagues, even as he rejected them. One widespread means for gaining influence was through prognostication about the course of a disease. The art of medical prognosis had a long history as a valuable service of the physician.[35] Yet physicians did not draw sharp distinctions between the goals of prognosis and the goals of treatment. Physicians used their statements about prognosis as support for therapeutic plans. The prognosis could, for example, be given in different versions to patient and to family, as a means of creating alliances to support treatment plans. Doctors, in fact, portrayed the use of prognostic statements as itself a form of treatment. These uses of prognosis as a support for, and an extension of, therapy are evident in the correspondence among Cabot's peers.

Writing confidentially to Cabot in 1912, a physician from Maine endorsed the idea that statements about his patient's condition were a crucial tool in managing treatment. In a letter referring the patient to Cabot's office, he filled in the background on what he had already said concerning prognosis: "I have simply told him that he had some trouble with his heart and kidneys." The doctor had softened his description of a condition that he actually thought was very grave. His patient was growing sicker, and it might be appropriate for Cabot to give him some additional warnings during his visit, he explained. The Maine doctor advised tact in this process, however. "I think it is just as well to let him down just as easy as we can," he continued to Cabot, "but tell what you think best in order that he may be better able to follow your instructions."[36] The goal was not to lay out the most accurate prognosis, but to phrase the medical opinion in a way that would enhance cooperation. Conveying just the right impression about the course of the disease might enable Cabot to gain compliance with his "instructions," and so greater therapeutic influence. Prognosis and therapy were mutually dependent actions.

It required a nuanced touch to achieve the right balance of prognostic impressions. While a sufficiently threatening condition might motivate a patient's careful attention to advice, too overwhelming a threat would inspire only resignation and retreat. Cabot's colleagues sometimes sought his assistance in achieving this desired balance. Dr. G. S. Foster wrote from Manchester, New Hampshire, in 1914 about his patient, a banker in his fifties, who suffered from the late stages of pernicious anemia. Foster was concerned that the banker was still hard at work, "settling some important estates etc," while his condition really required "to have him fully at rest both morally and physically." He was sending his patient down to Cabot's office for further advice. Perhaps, he suggested, Cabot could better convey to the man the seriousness of his problems in order "to make him feel that he cannot longer attend to these duties in any way." A proper warning might prompt the desired therapeutic change in behavior. Still, Dr. Foster worried that his patient was "failing very rapidly. " It would be important, he also advised, not to rob this man of the hope necessary to inspire continued attention to his health, so "we must withhold the hard truth."[37] A properly balanced prognosis about the disease would encourage the patient to make a therapeutic withdrawal from his business without abandoning further effort at appropriate care. The prognosis itself might in this way provide a therapeutic benefit.

These strategies for representing the disease to the patient were not a professional secret wielded unilaterally by physicians. The family of a patient often became allies and even accomplices in this process. The dynamics of disclosure about disease were evident to people who wrote to Cabot about the problem of cooperation with treatment. The relative or spouse who wished to inspire a sick person's commitment to therapy sometimes suggested collaboration with the doctor. Seeking to encourage his wife's willingness to treat her diabetes, one husband wrote asking Cabot to be prudent in discussing her prognosis. He explained that she had already been cautioned that she was *"threatened* with it [diabetes] and that she must be very careful of her diet." He was sure that by neglecting her diabetes, she was "acting in a way that will cause serious results." So cooperation with treatment was vital. However, she certainly should *not* know that she already had diabetes, because "if she had been told this it would have been fatal—it was hope that encouraged her to make an attempt to get strong."[38] This man also sought that delicate balance of a prognosis serious enough to inspire effort but not so grave as to instill despair.

Prognosis and Family Allies

Physicians in private practice had been accustomed to draw patients into treatment in part through the creation of such strategic alliances, not only against the disease, but also with the family and caretakers. The ability to influence adherence to treatment was the foundation of private practice. Most of the doctor's therapies in domestic settings were, after all, no more than injunctions to be carried out by others, to take an elixir, to alter the diet, or to monitor and respond with appropriate treatments. Doctors found their strongest allies in a patient's home, among the family who provided care and stayed with the person who was sick. This cooperation of the family might even serve better in some instances than the cooperation of the patient. "Better to leave your directions about medicine, food, etc., with the nurse, or whoever may be in charge, rather than the patient," advised one nineteenth-century physician discussing the management of the home visit.[39]

Cabot's colleagues tended to create complex alliances with family members around prognosis, giving differently weighted reports to patients and to their kin. A letter from a physician to a patient's brother, who forwarded it to Cabot, reported "a well established kidney degeneration I am sorry to say," while at the same time assuring the brother that the physician had not alarmed the patient himself "by laying too much stress on the kidney condition."[40] Physician and family should cooperate in controlling the patient's impressions about the disease. There seems to have been no one standard practice for disclosing dismaying opinions, and many of these communications display signs of improvisation. One of Cabot's patients reported having openly discussed the fact "that my trouble was positively fatal" with her physician, for example.[41] But an alliance with the family to protect the patient from bad news seems in general to have been desired, if not assumed. One woman wrote in 1914 asking whether Cabot suspected cancer in her husband, saying, "naturally we much rather he would not know."[42]

These practices in managing prognostic information imply certain assumptions about therapeutic influence, which are highlighted by Cabot's own opposition to them. Although he made important exceptions in his private practice, Cabot vigorously protested in general against withholding or manipulating diagnostic and prognostic information, and he actively undermined this practice among his peers on occasion.[43] His concerns about the legitimate basis of medical influence may have made him wary of manipulating information about disease. He often responded in a guarded way to the suggestion by other physicians that prognosis be used for therapeutic ends. The personal physician of one patient on whom Cabot had been consulted wrote hoping to inspire Cabot to make further efforts to stop this man's drinking. He asked Cabot "to bring upon [the patient] your influence in an endeavor to stop him from using alcoholic drink" by, for example, "explain[ing] the possibility of cirrhosis of the liver in cases addicted to alcoholic stimulation with a tendency to gout," and so forth. Cabot immediately penned a letter to this man about his drinking, ignoring the suggestion about cirrhosis and gout, since he had not diagnosed and did not prognosticate these diseases in this particular patient. Cabot also carefully disclosed in his letter that he was writing at the encouragement of the man's personal physician. After outlining some recurrent troubles with the pancreas that he thought might actually be a result of drinking, he explained that total abstention from alcohol was part of the treatment. Yet he allowed that "you may well think that the cure is worse than the disease. That's your business not mine. My only duty is to state the facts as clearly & honestly as I can."[44] Cabot felt a responsibility for the accuracy of information about disease that ruled out certain indirect therapeutic uses. He tended to be blunt in communicating news to his patients. Writing to a woman whom he diagnosed with circulatory problems, he reported, "I fear that it may trouble you somewhat for a good many years and I fear that medicine can give only partial relief."[45] His bluntness in diagnosis and prognosis was a source of harsh collegial criticism for Cabot, as he seemed to reject the valuable use of such information as a means of gaining the patient's cooperation.

For Cabot, the doctor's ability to identify and control disease was the primary justification for medical influence. Distorting or obscuring this information, even at the behest of the family, represented a breach of duty. While Cabot's call for physicians to be scrupulously honest in disclosing medical opinions still draws regular notice in present-day discussions, his efforts seem not to have affected actual contemporary practice much, although he did find a few medical allies in his day.[46] Despite the general opposition, his main argument seems incontrovertible, namely, that such deceptions would naturally become evident to the public, especially since they often relied on the collaboration of family members. A reputation for deception, Cabot sensibly maintained, tended to undermine the profession's broader credibility and authority. In essence, he argued that when doctors gave blatantly differing information to patients and to their families, they precluded recruiting the family as wholly trusting patients in the future.

Despite its pat logic, this argument found little support among Cabot's peers, who were reluctant to give up a practice that brought them valuable allies. They were, on the contrary, diligent in tallying up the bad effects that Cabot's policy of candid disclosure had on the patients exposed to it. The eminent cardiologist Paul Dudley White, who later became Cabot's own doctor, recalled the distress that Cabot's frank reports created among the patients whom they jointly cared for in the hospital, recounting how he had often had to return to calm and reassure patients whom his colleague had left alarmed by freely dispensed bad news.[47] Cabot's own brother Hugh was the source of a widely known story seeming to expose the foolishness of the former's frankness. Cabot, as Hugh told it, had found a suspicious growth on the cervix of a patient who consulted him and gave his patient to understand that she likely had a fatal cervical cancer. The woman reportedly left his office to quit her high-level job in order to prepare for her imminent death. She went next, however, to Hugh who was able to demonstrate through a surgical biopsy that the growth was actually harmless. Hugh reportedly never let his brother forget this, not so much for having been wrong, as for having been so painfully honest about his mistaken diagnosis.[48]

Tailoring a medical prognosis to support therapeutic advice was one means of gaining influence over patients, whose families were often willing partners in an effort to influence them through the careful management of prognostic information. In an office clinic like Cabot's, however, the traditional allies among family members might suddenly be absent. Inside the hospital, it was clearly the nurses who took over as allies of the physician's therapeutic control. The reorganization of the twentieth-century hospital and the creation of professional nursing only made these alliances stronger. A patient was likely to come to a medical office alone, however, without a companion or caretaker. An agreement about treatment had traditionally been struck with the patient, of course, but in the office, the patient became a more exclusive negotiator. Cabot's patients demonstrated an involvement with new forms of therapeutic influence based more on assessing and monitoring disease than on managing prognoses.

How to Know What Works

From the perspective of the person who was sick, the task of getting better could be perplexing. What evidence did one necessarily have either of the progress of health or of the effects of therapy? It was possible, of course, to reason based on how one felt and the nature of one's symptoms, especially in negotiating with one's physician. "I think that every thing is alright now as I don't feel them pains anymore," a Boston tailor wrote to Cabot in 1908, for example.[49] Another man sent a letter noting that he was coughing less, with less expectoration, and found that he was gaining weight; and so, he asked: "Question No. 1. Am I making progress?"[50] Some changes wrought by a therapy were conveniently evident to the eyes of the independent observer. One woman wrote that another doctor had recently "given me a remedy that has benefited me greatly and completely changed my color."[51] A change in complexion was the kind of evidence that other people could witness and confirm. A man who took Cabot's prescription for nitroglycerine reported that although he still had the same pains in his chest, "people that meet me say that I am looking a great deal better now" and that must count for something.[52] All sorts of evidence might confirm the benefits yielded by medical treatment. Yet patients still expressed apprehension about the gaps that arose between their physical appearance, what they felt, and the deeper alterations in health that they should perhaps be aiming for.

In the quest for therapy, such ambiguities made for difficult choices. The effort to get the right care could, for example, create difficult spirals, requiring people to continually adjust their treatments and assess the results. "[H]ow long should I take these drops?" an elderly man wanted to know. "I continue taking the 10 drops of medicine you prescribed for me three times a day." Although some of his troubles had subsided, he noticed that a new, disturbing symptom had arisen. Was this a sign of a new ailment, a new manifestation of the old condition, or simply "owing to the drops?"[53] It seemed difficult to be sure whether the medicine was slowly making him better or just substituting one trouble for another. Having received an apparently helpful remedy from Cabot, another man wrote back, still with some hesitancy, stating that after the treatment, "I was apparently as well as ever and do not know that I have ever felt any of the old symptoms." He seemed to be better but he expressed some uncertainty about the accuracy of his judgment on this matter. "I have just finished the 45 drops in the medicine which I have been taking," a stenographer wrote in 1912. "On the whole I think I am feeling somewhat better as I do not have the pain or pressure around my heart all of the time but after I work a little while it seems to come back."[54] Perhaps these patients were simply deferring to their physician by not overstepping their right to assess their own state of health. But they may also have been unsure about their ability to discern true improvement under medical treatments, or about what in fact constituted the best criteria for such progress.

Some patients, like the man who wrote to inquire about goat lymph and the woman who wanted to be treated with 606, hinted that their physicians should merely supply whatever it was that they deemed useful. Others adopted the passive role implied in a physiological therapeutics that relied exclusively on objective, measurable effects to guide treatment, in essence saying, "I shall do as you recommend."[55] Most people, however, seemed to accept a subtler and more involved relationship with their doctor. Even a gesture of unquestioning cooperation might serve more as a bargaining chip than a binding commitment. "I eagerly wait your prescribed regimen, which I suppose you will plan for me," one woman wrote. "I shall try to be faithful in carrying it out."[56] Her offer, which on the face of it looks like a promise of simple compliance, was in fact part of more extended negotiations in contracting for medical advice. The offer to carry out medical advice was an effort in soliciting sustained therapeutic counsel that she seems to have made before. She arrived in Cabot's office having already seen another physician, Dr. Agnes Victor, for the same set of problems. Dr. Victor had written to Cabot about her to complain that she "did not follow instructions for any length of time and after two weeks she did not return until December when she reported that she felt much worse in every way."[57] So her offer to follow the regimen Cabot prescribed was likely contingent on what happened next in this new therapeutic relationship.

The treatments that were available to these patients had both attractions and hazards that were vividly apparent from the very first days of Cabot's practice. After treating a young woman who came to him in 1898 with pain during urination, Cabot was able to celebrate in his office chart what he called a "brilliant therapeutic result." Prescribing ten grains of a chemical called urotrophine for her to take three times a day, he noted in her chart on a next visit that her problems had "gradually decreased" and the "urine (in about 5 days) cleared," with confirmation obtained by the repeated microscopic examination of her urine.[58] Similar therapeutic satisfactions were evident to another physician who wrote to Cabot in 1919 pleased about results for a man that he had treated for "streptococci [bacteria] growing in his blood." "I gave him mercury bichloride intravenously and he promptly fully recovered," the doctor announced, concluding with the definitive evidence that "his blood was sterile on culture."[59] In these brief success stories, Cabot and his peers sketched the outline of a powerful therapeutics. People with clearly defined disturbances in the blood or urine received exact chemical treatments followed by physiological testing to confirm the effects. If Cabot and his colleagues hoped to deploy this simple logic to recruit the support of their patients in daily practice, however, they faced several significant obstacles. Uncomplicated cures and immediate relief were rare. The same woman whom he had cured of urinary troubles returned to Cabot's office the next year with the same troubles. She was treated again with urotrophine, and the result noted in the chart this time was: "miscarriage and transfer to Camb[ridge] MD."[60] A treatment that seemed demonstrably effective against disease might have other distressing consequences.

These treatments were often highly potent and alluring in their promise, but equally dangerous and difficult to apply. Many novel treatments were emerging from the pharmaceutical laboratory and the surgical theater that, in the words of the New York physician Samuel Meltzer, "enabled us in some instances to grapple with the disease itself,"[61] a struggle that the patient at least might consider with a certain ambivalence. The example that Meltzer found close at hand in 1911 would become a familiar one for Cabot's patients. That year marked the announcement of the miraculous drug 606, "an efficient, specific, synthetic drug scientifically developed" for the treatment of syphilis.[62] Salvarsan carried great promise, but also imposed tough choices on those who would employ it. Surveying the results from its early use in the pages of *Popular Science Monthly*, Dr. Fielding Garrison warned that twelve deaths among the first thousands of cases, along with a course of treatment that was "exceedingly painful in the first stages," had led the inventor of the drug himself to "compare it with operative surgery in that it can never be given without certain risks."[63] Salvarsan "marks an epoch in medicine," said Meltzer, suggesting that it figured as part of a general change under way in the doctor's therapeutic armament.[64]

The Value of One Patient's Treatment

Some patients seemed surprisingly appreciative of their doctor's control over physiological manifestations of disease, even when it served their expressed interests only indirectly. Physiological therapy, at least in Cabot's clinic, offered them often weak and partial solutions to their troubles. Yet patients sometimes endured long trials with such treatment. The individual medical practice was a good site for recruiting and educating patients, so that a regimen of physiological monitoring could be established as the form of care. One thing that this care offered was protection, or at least support, in the patient's perplexing choices about treatment. While physiological treatment rarely offered the cheap, sure, and rapid benefits demanded by the crude consumers parodied in the medical literature, it displayed an impressive control over the entities that it defined. For the patients who accepted the premises, physiological treatments did offer a proof of control that could be perceived as a service in its own right. One example helps to demonstrate how patients could accept the physician's markers of progress as their own. Some patients not only identified an independent value in the doctor's control over disease, but seemed capable of sharing it vicariously.

In May 1902 a middle-aged woman, whom I shall call Ella Watson, came to Cabot with a complicated story about many ailments and treatments. She had "pain with urination" and severe pain at the waist when she bent to sit, as though, she said, she were being folded up "like an accordion." Cabot reported that she had been "purging and vomiting," which likely meant taking medications to produce these effects. She had tried "rectal injections" and had "had bladder washings done." Some of the therapy seemed to help. Previous treatments, he noted in the chart, "have made her able to sit." Reflecting back on the months before her visit, she herself noted, "I am better now than I was . . . I can walk now it hurts me much less." She was still suffering, however, and looking for further guidance. She had come to Cabot's office, she reflected in a subsequent letter, because "I believe . . . if any one can assist me to health you can."[65]

Assisting to health might involve many things. If it meant medical treatment, then Cabot insisted on naming the disease to be treated. During their first meeting, in addition to the usual detailed interviewing and examination, he subjected Watson's urine to several chemical and microscopic tests, which made evident certain worrisome abnormalities there. Perhaps he discussed the finding with Watson. Following this visit, she and Cabot both fixed their attention more determinedly on the urine. Watson went home and returned for a next visit with dutiful notes detailing her experiences with urination over a couple of days. She charted observations on when it occurred, how much, and how it felt to urinate after rising from different positions. On a separate sheet of paper she recorded the times of urination over the course of an entire night and carefully described some solid material that she had passed, including a small sketch of it in the margin of her notes.[66] She returned with this evidence and thirteen separate samples of urine that she had collected. Cabot reciprocated by analyzing the samples and confirming that there was indeed a problem in the urine; and he offered some new advice.[67]

With their mutual concern about the urine secured, Cabot recommended that Watson next see a surgical colleague who specialized in the treatment of the bladder and kidney. The surgeon, Dr. Edward Reynolds, could employ a special lighted tube called a cystoscope to examine the inside of her bladder and possibly treat the problems there directly. But this recommendation did not at first win Watson's support. "I dread an instrument examination," she wrote back to Cabot after their second meeting.

She had good cause, as she explained in the letter: "a Somerville lady was dilated by Dr. Morris Richardson [a surgical colleague of Reynolds's]" she wrote, "and could not hold her urine afterward." She worried about the treatments that a surgeon might apply, writing that "those physicians who have instruments and the knife I am afraid of."[68] Perhaps it made some sense, however, to pursue the trouble where it seemed to reside, in the urine and bladder. A short time after writing about her anxieties over surgeons, she did go to visit Reynolds in his office just down the block from Cabot on Marlborough Street.

Over the next three years, in a series of letters from his office, Reynolds reported back to Cabot about Watson's ongoing treatment. He used the cystoscope, as anticipated, to examine the interior of her bladder and found a specific source for the continuing troubles. Testing her urine on a guinea pig, he confirmed what his examination had suggested, that she had a tubercular infection of the urinary system. With this diagnosis established, Reynolds persuaded Watson next to come to the Deaconess Hospital, located at this time in a couple of adjacent row houses on Massachusetts Avenue, not far from Reynolds's office.[69] There, Reynolds used a more involved technique with the cystoscope again to sample the urine separately from each kidney through the two ureters that emptied into the bladder. No surgery took place; but with the tuberculosis effectively localized to only one kidney a surgical solution presented itself. One kidney was infected while the other kidney tested as normal during this sampling procedure. Reynolds managed to have Watson enter Deaconess Hospital again. In a subsequent publication outlining this lengthy case, Reynolds described a lengthy surgery that he performed on Watson at the Deaconess, where he removed her infected kidney, leaving the healthy kidney intact.[70] Following the surgery, Reynolds wrote back to Cabot that Watson had gained back much of her weight. Her original urinary symptoms persisted, he confided, but were certainly attributable to localized areas of continued infection now in the bladder; and went on to report that these lesions were, in fact, visible on a repeat examination with the cystoscope.[71]

For another year, Reynolds continued his efforts to eradicate from Watson's bladder any testable evidence of tuberculosis reporting his progress in an ongoing correspondence with Cabot. He performed repeated examinations with the cystoscope, applying silver nitrate and heat to suspicious areas in Watson's bladder, followed by confirmatory testing for tuberculosis. After a long course of treatment and monitoring Reynolds was able to boast finally in 1905 of two successive negative tests on Watson's urine. It was an excellent example, he wrote to Cabot, of what could be accomplished through "determination to see the case through on the part of both patient and attendant."[72] In his published account of this surgery, Reynolds made it clear that the primary accomplishment of his work with Watson lay in the eradication of the testable evidence of tuberculosis. He concluded his published account of the case by noting incidentally that "this patient still suffers somewhat from a contracted bladder, due undoubtedly to long-continued inflammation."[73] The urine, however, remained entirely clear of tuberculosis.

We might wonder at Watson's own impressions about the utility of a treatment whose primary outcome was normal tests. The sources of her determination to see through several years' treatments with "instruments and the knife" remain obscure. Watson did, however, manage to leave behind a personal summary of her long experience under Reynolds's care. Enclosed in her office chart is a final letter to Cabot dated 1908. Trying to sum up the three years of surgical treatments, she remained tentative. "I think I am a well woman," she wrote. She had, she ventured, some "little trouble with bladder now . . . Dr. Reynolds says I will have trouble with that as long as I live in extremes of heat or cold weather as I have thus far."[74] She and Reynolds agreed on the incomplete resolution of her urinary symptoms. She did "still suffer somewhat," as he had reported in the article, but it was after all a difficult world, full of inclement weather.

In this final letter, Watson was able to point to one certifiable benefit of her determination to see her treatments through to their conclusion. Dr. Reynolds had claimed that his treatments corrected an abnormality in her urine, eliminating all traces of tuberculosis. Such monitoring of the urine had been part of Watson's first exchanges with Cabot when she originally sought his assistance. So in concluding, Watson recalled for Cabot a last report on the urine that she had received from Reynolds following the completion of her treatments. He had analyzed a final sample of urine and written a reply that she now cited back to Cabot. "The specimen of urine which you left me," she quoted from Reynolds's note, "appears to me to be about as normal as anything could well be—I know that will please you." She ended her own assessment of the treatments with this same material assurance that had begun it. Reporting to Cabot about her health, she turned again to the evidence of the urine, which had been a basis for their mutual planning. The problem with her urine had been corrected. She would be pleased about the demonstrably normal tests, Reynolds had asserted. And perhaps she was. She closed her letter, "I am very gratefully yours, Ella A. Watson."[75]

Cautious Consumers of Treatment

The patients who wrote to Cabot did not in general protest that they needed something fast, cheap, or sure from him or his colleagues. They did deliberate a great deal about their acceptance of their doctor's advice and prescriptions. Some patients characterized their acceptance of treatment as a cautious step, for which they took responsibility. Watson had written to Cabot soon after her second visit, questioning his recommendation of surgery. Weighing her choices, she reminded him that "I want to do what is best for me"—as though admitting that she might finally have to be the judge of that.[76] When patients felt that they had chosen correctly, such a responsibility seemed easier to bear. Another patient, an office clerk from a nearby town, wrote to say that he had decided not to go on with Cabot's prescribed therapy, stating, "I have discontinued the treatment allowing [the rheumatism] to takes its own course." "The result," the clerk informed his doctor, "is that it has left me entirely."[77] He had evidently made a good choice.

Patients sometimes regretted having forgone treatment, however, especially when the untreated condition did not improve. "[I]f I could follow your advice of a change of climate and associations of two to three years I might gain in strength—I cannot tell," one man wrote Cabot.[78] "If I could have acted upon your advice I would not have much trouble," a second patient similarly wrote. However, a third reported: "I followed your advice . . . and took the pills and digestive biscuits," but without getting much better.[79] If patients sought and considered therapy, when should they then reject it, and on what grounds?

One young man, whom Cabot described only as a "baggage recorder," wrote reporting that he had not had much relief since his last visit, and in fact felt worse after taking the doctor's treatments. He was not giving up on the treatments yet, however. "I shall continue a little longer," he conceded, "with medicine and 'rules' as near as possible . . . and may yet get [the] 'desired result.' "[80] It was sometimes hard to know what to expect of a treatment or medication. Patients expressed complex understandings of the desired effects. A young man wrote about his medication, for example: "I think the first two or three days after taking [it] I had more distress in my stomach . . . and I was somewhat pleased for I thought it was going to have a different effect after the newness of the change was over for a great many times a medicine that helps often upsets one at first."[81] Symptoms were changeable and difficult to convey to others. They might seem an insufficient basis for tough choices about a lengthy, risky, or discomforting treatment, especially when the ultimate goal was the eradication of a disease.

The concrete facts of physiologic monitoring offered a shared territory lying between the patient's unimpeachable, if inaccessible, claims about symptoms and the physician's assumed expertise. Perhaps Ella Watson had difficulty assessing how her bladder problems in 1908 differed from her problems in 1902, after years of treatments. She could in any case offer the evidence of her physician's confirmatory testing of the urine. Such evidence also provided a reasonable leverage in the process of understanding and negotiating treatment. This use proved attractive to a stenographer in her fifties who wrote in 1915 about the strict dietary recommendations that Cabot had made for controlling her diabetes. Body weight provided one important piece of information about the course of her disease: "[T]he fact that my weight is now 184 1/4 pounds on bathroom scales (no encumbrances) may not be of thrilling interest to you, but I tell you this because I want some fruit," she wrote.[82] If part of the authority over eating fruit did lie with her doctor, it was at least an authority responsive to petition. Measured weight was a criterion more equally accessible to doctor and patient than were either the symptoms or the process of disease. Such shared criteria could prove useful in negotiating plans about treatment. "I feel good," on the other hand, might seem as much a pleasantry as it did a basis for seeking to influence the physician's advice.

Patients were sensitive to some odd discrepancies in the results of medical treatment. If symptoms posed a challenge to the physician, then they might also pose a challenge to the patient hoping to gain the advantages of specific therapies. "I think the Treatments help[ed?] me a great deal," wrote one perplexed patient, "but I came home in just about as much pain as ever but I gave up all medicine and keep Cheerful."[83] Perhaps this man was merely confused about the purposes of his treatment, but he may also have acknowledged an imperfect relationship between how he felt and what his treatments aimed to accomplish. Similarly, the statement cited earlier that "the operation itself was a success, but I have never recovered from the check to my nervous system,"[84] might have come from a patient who recognized the paradox involved. It seems an earnest attempt to note the evident tensions between therapeutic goals and subjective good health.

Cabot's patients sometimes demonstrated an appreciation of physiological monitoring of therapies, when it was available. One middle-aged man, a minister, developed explicit criteria for his disease and medical care using specialized testing. He came to Cabot in July 1903. Evidence from their interview, a physical examination, and microscopic examination of his blood yielded the unfortunate diagnosis of pernicious anemia. The disease had a difficult reputation at that time. Although potentially fatal, its course could be unpredictably mild and prolonged. It might leave a person on his feet for years or quickly create devastating troubles.[85] The minister wrote to Cabot over the next year in ongoing efforts to track the treatment and control of this shifty malady. Physiological monitoring of his blood played a central role.

A couple of months after his first visit, he wrote back from New Bedford, where he had met with another physician. Another blood examination by this second physician had found red cell counts slightly different from Cabot's. The New Bedford physician, Dr. Connor, had prescribed a new medicine for him, he reported. It raised for him an important question: "Do *you* think I should take the medicine?" In providing Cabot for a basis on which to make his recommendation, he noted, "I find that in three months [since our visit] I have gained in number of red cells from 1,000,000 to 3,480,000."[86] He also noted that he was feeling better, evidence well supported in the rising number of red blood cells. Did he really need a change in medication as Connor recommended? The question as it was posed in the minister's correspondence seems as much directed to the minister himself as to his physicians.

In February of the next year, he wrote to Cabot asking further advice about treatment, this time less sanguine about the results of monitoring. He had been back to Dr. Connor earlier in December for further monitoring. "I found that I had gone back to 2,546,000," he reported. "I fear the month of January has taken me down still more." He described weakness and malaise, which seemed to confirm the information from the blood counts. He inquired about other possible therapies that he might pursue, asking: "I have an electric battery, continuous current, but do not use it. Could it be made to do me any good?"[87] He drew his physicians into the deliberations over treatments using the information of blood counts as their shared evidence about the disease. He last checked in again about seven months later, writing from Nebraska to say that he hoped to be in Omaha soon, where he believed he could obtain another blood count, which he would forward on to Cabot.[88]

Falsifying the Value of Treatment

Both the minister and Ella Watson had in their own ways turned to the evidence of laboratory testing to judge and communicate the results of their medical care. Protracted illnesses like those associated with urinary tuberculosis or pernicious anemia might wax and wane under a great variety of influences, including treatment. The blood count or the urinary bacteria seemed to represent acceptable targets for therapy that might be drawn out over years. If improvements in the tests did not always correlate with a marked sense of improvement in health, at least they served as a reasonable means of communicating with their physicians about matters of mutual concern.

Physiological monitoring provide one means to confirm the value of a medical treatment when the way it made you feel seemed insufficiently persuasive evidence. Monitoring might also serve to expose a physician who tried to foist an inadequate treatment on a patient when the latter's testimony alone could not indict him, as in the following example. The husband of a fifty-year-old woman who had been a patient of Cabot's wrote to the doctor in 1915 to inform him of her recent death and of her mishandling by another of her physicians. She had first come to Cabot's office in 1912 suffering from weakness, weight loss, and stomach pains. Cabot examined her blood during this visit and confirmed the diagnosis of pernicious anemia that she had received from her doctor at home. Although this agreement on the condition by independent doctors in different states may not have been especially reassuring at the time, it seemed to the husband worthy of note in light of what followed. After the visit to the Marlborough Street office in Boston, she had returned to her doctor at home, Dr. Fulton, where she continued with a conventional line of treatment that he had begun. Over the next year, several laboratory reports on her blood reached Cabot from Fulton's office and were dutifully filed away, diagramming her ongoing struggle with the disease. Then in 1914, Fulton wrote about an important development. His patient had heard of a new treatment with the radioactive element thorium being used in Berlin and, being eager to try it, she was traveling to Germany to put herself under the care of an expert at the university there, Dr. Bickel.

Her husband then took up the account in his subsequent letter, reporting to Cabot that "Dr. Bickel acted the part of a faker." Bickel had provided his wife with the thorium treatments and claimed to demonstrate their powerful effectiveness against her disease through a series of gradually improving blood tests that he showed to her. In fact, after a short time these tests seemed to display remarkably normal results for her blood. Apparently skeptical, however, his wife had sought out a second opinion from another physician in a nearby city. Her husband continued the story, "within a few hours after leaving Berlin she had a thorough examination by Dr. Von Noorden and he found her blood count nothing at all like Bickel had reported it." In fact, her blood tests still demonstrated exactly the signs of the pernicious anemia that had plagued her all along.[89] The whole sad tale of these false treatments was spelled out in the physiological monitoring of her blood, which also allowed her husband to provide creditable evidence of how she was misled by this one physician. He concluded the story to Cabot, noting that his wife had died shortly after her return home from Germany. The results of blood tests remained her husband's best evidence for what had happened.

Placebos and the Authority to Treat

The use of physiological monitoring of treatment with blood counts or urine tests offered independent support for the patient's trust and cooperation; but such use did not supplant other, more personal forms of medical influence. Early twentieth-century physicians continued to weave therapeutic intentions into their discussions of diagnosis and prognosis, as we saw earlier. They did not in general abandon their claim that the authority to treat derived as much from a privileged understanding of the individual patient's circumstances as from an objective knowledge of the disease.

Cabot himself argued that new physiologically guided treatment remained, like all good therapeutics, "based on the individual's manner of reacting to the disease . . . [accounting for] the whole individual so far as it [therapeutics] can discover him."[90] Such claims, however, took on changed significance in the context of new medications and treatments. Treatments that targeted specific diseases, like 606, abdominal surgery, and thorium, promised a future filled with cures that were quick, sure, and easy. In actual use, however, they required patience, sacrifice, and compromise. The emerging tension between the promises of modern therapy and its applications pushed physicians to reconfigure certain traditional bases of authority. Debate about placebo treatment among Cabot and his peers, although a minor skirmish, reflected and sharpened these concerns over the sources and justification of therapeutic influence.

Placebo medications such as a vial of colored water or a sugar pill had no defined therapeutic action that could be monitored or assessed. The doctor who offered a placebo was making the gesture of treatment without its expected content. Placebo treatment thus exposed questions about the basis of medical authority in a stark form. Cabot was an especially opinionated party in the debate over placebos. His arguments still find frequent reference in present-day discussions of the ethics of placebo use, but are equally valuable perhaps as an insight into the issues of his day.[91] The placebo, Cabot claimed, was nothing more than a lie about therapy, and it was thus unacceptable in practice.[92] As a treatment devoid of specific effects, the placebo placed the doctor's control over treatment at stake without its usual justification in technical knowledge about the diseased body.

It is important to recognize that the use and significance of the placebo were simpler for Cabot and his peers than they would become subsequently. By the mid twentieth century, placebos had developed into valuable tools of medical research, serving as a baseline or a control against which researchers could measure the effects of other treatments. By the late twentieth century, medical researchers were increasingly being pressured to draw sharp distinctions between conventional and experimental therapy, and between experimental therapy and placebo-controlled experimentation. The use of placebos as a control in medical research became emblematic of the fundamental conflict of interest between researchers and research subjects. The intention to treat differed crucially from the intention to experiment. The use of placebos in experiment seemed to make this fact abundantly clear. For physicians like Cabot in the early twentieth century, however, placebos remained just a particular form of treatment. The concept of a well-defined "placebo effect" of importance to the clinical evaluation of novel treatments still lay several decades in the future. Yet placebos raised equally pressing questions for Cabot and his peers about the nature of medical authority over treatment.[93]

Cabot warned that the use of placebos imperiled the reputation of physicians broadly. Although his criticisms of placebo treatment seemed an application of his concern about separating medical authority from its technical justifications in the management of disease, they also spoke also to broader issues. In attacking placebos, he mirrored the pragmatic argument that he had made against the conscious manipulation of prognostic or diagnostic information. A patient who found that a physician was dispensing placebos might well ask, "What other tricks will he think it best to play on me for my own good in future?"[94] People might come to doubt or challenge the doctor's therapeutic advice, just as they would doubt prognostic statements once having found out that they might be altered to suit ulterior purposes. Rational persuasion was a cornerstone in Cabot's conception of the doctor's authority. To recruit a patient into treatment, Cabot advised, the first steps were to "take the patient into our confidence. . . . tell him the truth, explain his malady, and the means of its cure. "[95] Placebos might limit medicine's ability to draw upon public trust and credence in such dealings. In addition, the use of placebos would risk separating individual therapeutic influence over patients from its material justification. If therapeutic authority was grounded in special technical knowledge about the diseased body, then the use of placebos seemed a willful misrepresentation of this authority.

That the use of placebos would require professional discretion was evident to physicians long before Cabot's critique. Until the mid twentieth century, when they took on a central role as research tools, placebos were seldom discussed in the medical literature.[96] The use of placebos required an element of secrecy almost by definition; and its mention was often accompanied by advice to be circumspect. An early nineteenth-century description of the placebo in a *Medical Lexicon* called it a substance "intended rather to please a patient than to cure a disease."[97] But what was pleasing about a treatment with an ordinary, inert substance like sugar unless its true nature remained hidden? It would be paradoxical to champion placebos publicly, as open discussion tended almost necessarily to undermine their use. Discussion of the tacit use of inert therapies was thus buried in medical communications intended for professional audiences.[98] To speak publicly about placebos was to eschew them.

Although this secrecy has made them difficult to track, placebos were by all evidence widely used in medical practice throughout the nineteenth century and into the twentieth.[99] Cabot documented his own explicit use of placebos in his private practice before he came to reject them around 1903.[100] The placebo was one application of a general tenet of therapeutics that long predated the introduction of this specific term into medical parlance. In eighteenth-century medicine, a positive influence over the patient's understandings and beliefs figured as an indispensable part of practice. Enlightenment physicians argued that the patient's imagination was a valuable avenue for exerting a physician's therapeutic influence. By the end of the long eighteenth century, physicians were linking such influence directly to the intentional use of placebos to create favorable and hopeful impressions. By the end of the nineteenth century, a similar rationale had developed for the use of the placebo as an adjunct to mental suggestion. The placebo looked like the concrete form of a therapeutic suggestion: the suggestion that effective treatment was being provided. Placebos found relatively uncontroversial use in the nineteenth century as a therapeutic tool not too different in practice from the physician's established and accepted resources of reassurance, support, and encouragement.[101]

Placebos and the Lazy Patient

Long familiarity with the use of inert treatments in this tradition perhaps buffered Cabot's contemporaries against his dire warnings. Few at the time went as far as he did in renouncing the placebo as illegitimate or dangerous to medical reputation. Instead, those who criticized the placebo echoed familiar concerns about giving in to the weaknesses of patients. Placebos, they cautioned, pandered to the patient's desire for an easy solution. Francis Peabody made light of the physician who indulged a patient's anxieties through "cheerful reassurance combined with a placebo."[102] In 1938, William Houston, a Texas internist and member of the prestigious American College of Physicians, aired a fascinating and extended argument about placebos and the legitimate therapeutic influence of the physician. In his account, placebos endangered the doctor's efforts "to educate his public to demand something different, something far better," such as full explanations and truthful advice.[103] The principal harm done by the placebo was inspiring medical laziness, Houston said. Physicians tended to reach for placebos to treat minor symptoms in the absence of severe diseases. Placebos thus both substituted for the physician's earnest efforts to explain why no medical treatments were indicated and indulged patients by teaching them "to expect a medicine for every symptom."[104] Physicians should fight against the temptation to take the easy route in prescribing placebos. It might also undermine their ability to pursue more complex and taxing therapeutics when they were required. Patients who grew accustomed to the relatively easy task of taking an (inert) pill for every ache would balk before the difficulties of applying a more taxing treatment to address the fundamental cause of a disease.

Other physicians, who identified a similar risk in indulging patients with placebos, continued to affirm a noncontroversial role for them as an element of treatment compatible with medical authority. For George Roland, writing in 1908, placebos were minor aids to the doctor's therapeutic influence over a patient. Affirming the traditional idea of the therapeutic potential to be found in all interactions with patients, Roland argued that prescribing pills of any kind in fact distracted from more fundamental therapeutic obligations. Pills were also the easier path, since "patients will always take medicine when they will not take advice." So it was "the physician whose force of character makes his advice sought after and followed . . . who accomplishes the most good," rather than the doctor who gave out pills, placebos or otherwise. Physicians achieved their greatest gains by influencing "the correction of the habits and customs of the patient," enforcing moderate diet, outdoor activity, regular routines and schedule, as well as the proper use of medications. However, since patients often sought out the doctor primarily to get a prescription, it was reasonable and sometimes helpful in Roland's thinking to give placebos. He argued that in some cases "a teaspoonful of colored sweetened water every two hours will come as near meeting all the requirements as any of all the multiplicity of medicines recommended."[105] If an inert placebo treatment supported the doctor's beneficial therapeutic influence, then it counted as legitimate.

To Roland, a placebo medication threatened no special compromise or betrayal of professional duty. Placebos were just another tool in the doctor's black bag, valuable to the degree that they helped to encourage much-needed attention to personal habits and hygiene. Like his nineteenth-century predecessors, Roland sought to identify therapeutic purpose in all interactions with his patients. Anything that supported a patient's attention to a healthy way of life counted as a valid medical treatment. Once again the fickleness of patients seemed to present the greatest challenge to the doctor's ability to recruit to treatment. The major task facing the early twentieth-century physician, to Roland's way of thinking, was in asserting a strong therapeutic influence— rather than in attempting to justify such influence, as Cabot would argue.

The Placebo as Psychological Influence

Despite the willingness of physicians like Roland and Houston to minimize the problems of placebo use, there were rifts opening during this period in professional agreement about placebos over and above the controversies outlined by Cabot. Unusual or controversial placebo treatments sometimes touched off intense dispute and debate. Physicians who followed the medical literature in 1918 would have been struck by the remarkable case of Dr. Hildred Carlill in the pages of the British medical journal *Lancet*. Carlill provoked a spirited debate by publishing a defense of his placebo treatment of a patient who suffered from sudden debilitating attacks of sleepiness. He reported a complete cure of this ailing young man through an elaborately contrived, and risky, sham surgical procedure that involved removing a small section of skull bone. He reported that although there was no defined, physiological rationale for this procedure, it had produced favorable results by creating the psychological impression that it was a beneficial intervention. The treatment represented the successful use of a placebo surgery to cure a patient of a hazardous condition, Carlill argued. He attempted to justify his treatment as a valid medical action, intended only to benefit his patient. He understood his individual patient well, he explained, and had researched his disorder carefully; he had a duty to act upon this potentially remediable problem when he felt that he had a probable solution to offer. The sham surgical procedure was his best attempt to remedy his patient's difficulty.

Other physicians wrote to the *Lancet* joining in on his side. "Dr. Carlill," argued one supporter, "was actuated solely by the desire to benefit a patient, who, as long as he remained uncured, was in danger."[106] In replying to the first raft of criticisms, Carlill himself went so far as to claim that any form of treatment administered by a physician was acceptable, "so long as there is a remote chance that it will benefit the patient."[107] After all, Carlill asserted, the patient "prefers a cure, however produced."[108] Carlill suggested that the placebo was simply another version of what the patient sought from him and what he had the duty to provide, a potentially helpful treatment.

Carlill's opponents too seemed to accept the premise that a placebo treatment was one way that a doctor could respond to the patient's uncomplicated request for help. None of the respondents in the *Lancet* attacked placebos simply on the basis that they were deceptive, as Cabot had. Instead, they raised questions about the implications of placebo treatments for more general therapeutic influence. Adopting a familiar tone of wariness about the average patient's appreciation of the challenges of therapeutic compliance, Carlill's opponents warned that placebos were yet another form of medical pandering. One opponent, Dr. Ready, took up a line of criticism similar to that of Peabody, Roland, and Houston, suggesting that placebo treatments indulged the patient with the promise of simple solutions and easy relief. Dr. Ready asked whether even successful placebo treatments "are good or justified when no effort is made by the patient and nothing is given to dethrone his egoism."[109] Placebos gave in to laziness on the part of both physicians and patients in reaching agreement about more difficult and complex therapeutic plans.

We can also hear in Dr. Ready's reply in 1918 the hint of emerging ideas about a vigorous new branch of therapeutics. His suggestion to "dethrone egoism" reflected a psychodynamic understanding of interactions with patients being pursued most concertedly in the growing medical literature on Freudian psychoanalysis.[110] Perhaps the doctor's personal relationship with patients should be distinguished sharply from conventional medical therapeutics, so that advising about prognosis had no legitimate independent role as a form of treatment. Nonetheless, there might be purposeful, therapeutic uses for the medical relationship. Houston, in his 1938 article "The Doctor Himself as a Therapeutic Agent," and in a longer prior monograph, explored this notion in greater detail.[111] Houston asked whether physicians could anticipate and control the therapeutic effects of the doctor's personal connection to the patient. He reiterated Francis

Peabody's notion that there were healing bonds formed in the relationship between physicians and patients, validating the therapeutic nature of personal interactions. Drawing in part from psychodynamic teachings of the day, Houston asked, "[H]ow can the doctor himself, as a therapeutic agent, be refined and polished to make of him a more potent agent?"[112] Placebos were to Houston no significant threat to therapeutic authority, as was shown earlier. They were only distractions from a more thoughtful application of related modes of influence.[113] Freudian psychodynamics aspired to affect mental aberration through the medium of the therapist's relationship to the patient, using the relationship as a tool for detection and treatment. Houston, and to a lesser extent Peabody too, assumed the existence of such therapeutic relationships. With their focus on physicians in general medical practice, however, they took as their goal, not the treatment of mental pathology, but the creation of general medical therapeutic influence. They took an instrumental view of the medical relationship that paralleled the new psychodynamic model and reconfigured the doctor's traditional therapeutic interpretation of efforts at reassurance, support, and encouragement.[114]

Cabot and others who debated the use of placebos often characterized the sugar pill as a simple entity; for Cabot, it was nothing but a deception. But placebos were more potent and complex mixtures than was allowed in these critiques, as the psychodynamic interpretations of Houston and others would suggest. Isolating the placebo so cleanly from its manifold meanings and associations was not easy. Cabot himself had organized and tucked away evidence about the extended significance of placebos that might have given him pause. One person who wrote asking him to soften his opposition to the sugar pill brought a complicated personal history to bear on the issue. She had never been a patient of Cabot's but knew him socially and maintained a curious association over several years. Her first two letters, which Cabot filed away with his personal correspondence, sought out his company at an evening social engagement and then subsequently extended another invitation, with the reassurance that his previous abbreviated attendance had not been misconstrued.

In a series of personal letters over the following year, she continued what seems like a determined flirtation in the face of a lukewarm reception. She wrote to marvel that Cabot's companionship was to her "more like that of a brother's than any I ever expected to have" and described her embarrassment at finding missing buttons on the front of her coat after an evening out together. She inquired in detail about his understanding of the commandment to love thy neighbor and wrote again to excuse him from his absence at another event, where she had been anxious to see him. Then she wrote to mention that she would be taking a walk to a particular bridge at a certain time and day and said she would be glad if he happened to be free to meet her there.[115] She assured him that this was her planned itinerary and should he not be able to make it, he would be excused in advance for his absence. Following this letter, there was a long break in the correspondence.

Her final letter in the collection turned to the subject of placebos. She had just read Cabot's denunciation of the medical use of placebos and wrote to ask whether they might not in fact be a good thing sometimes. Honesty on the doctor's part was important, she agreed. But could he not allow "that in highly nervous cases and in unbalanced minds where reasoning is practically impossible, a placebo may be necessary?"[116] How could withholding care in such cases be better? She concluded by requesting copies of his paper for two physicians of her acquaintance. Perhaps in this way she might be able to gain some concession from the good doctor, who proposed to give not even the semblance of treatment to certain patients. A placebo was at least a substantive response to the request for assistance and must have seemed better to her in some ways than the honest response that there was nothing to give.

PLACEBOS SEEMED TO CABOT to present a peril to the physician's ability to guide treatment. He joined a larger debate among his contemporaries that explored the changing basis of therapeutic authority. Placebos were a minor concern but pointed to larger tensions developing between the doctor's therapeutic influence and its basis in the verifiable control of disease, at a time when doctors increasingly aspired to such control. By the mid twentieth century, placebo effects would become a common baseline serving to define and measure treatments. Valid medical treatments became by one definition interventions that were better than the placebo. But in Cabot's time, placebo treatments still squared nicely with the use of prognostic reassurance, admonishment, or encouragement, and other well-established means of supporting and enforcing therapy. Physicians had long claimed that their power to treat derived from a general obligation to help the sick, from a privileged understanding of the individual patient's constitution and needs, and from a willingness to put the patient's interests foremost. But in the isolated realm of the office clinic, in the face of increasingly high-stakes therapies, would such justifications continue to suffice? Physicians like Houston responded in part by trying to redefine a therapeutic role for the medical relationship in psychodynamic terms, independent of the technical services delivered in the treatment of disease, but potentially supportive of them.

The challenges facing the doctor's therapeutic influence had also changed. Cabot and his peers expressed concern that patients would not tolerate the demands of physiological treatment. They aired their concerns in professional communications that parodied the impatient consumer of medical services searching for a quick and easy fix. Similar discussions spilled over into the debate on placebos. Placebos indulged the patient's desire for a pill for each and every ill, Cabot warned. In contrast, careful, physiologically guided therapeutics required patients to accept treatments that sometimes sought obscure targets, like red-cell counts, and created results that were not immediately evident, except through technical methods. Physicians inspired by these efforts asserted the need to control treatment in a way that was independent of the reports of symptoms and effects by patients. Physiological monitoring of therapy could both confirm the effects of treatments and create the framework for the ongoing supervision of treatment. Paradoxically, treatment that was quick and simple also seemed to some doctors to be the most detrimental to the physician's authority. The professional literature of the early twentieth century warned against the hazards of a market filled with easily obtained patent medicines, panaceas, and quack cures.

The actual give-and-take of negotiations over treatment in Cabot's clinic reveals a much greater variety and complexity of interaction than was imputed by the medical literature. People writing to Cabot rarely articulated an interest in treatments that were quick, cheap, and certain, however desirable they might be in the abstract. Perhaps they understood that it was better to seek such services elsewhere, despite the apparent promise of modern medical science. Patients proved flexible in petitioning for Cabot's services and demonstrate concerns in their correspondence that differ strikingly from what the medical literature attributed to them. If their physicians seemed wary of the subjectivity of symptoms, patients replicated this concern, often noting in their correspondence that their symptoms gave them difficult and sometimes conflicting evidence. What did it mean to get better from treatment? Illnesses waxed and waned and followed their own unpredictable courses. The response to therapy might be ambiguous and impossible to distinguish from the natural changes that typically accompanied illness. Patients accepted medical reassurance about the validity of medical therapy through physiological monitoring to the extent even of discounting their direct experience. Physicians like Fulton, Von Noorden, Reynolds, and Cabot attempted to make their clinical value demonstrable in the physiological monitoring of disease. One patient, Ella Watson, even suggested subtly that her physician's ability to chart and document control over a disease had its own vicarious value for someone who was vexed with a chronic illness.

CHAPTER 5

NERVOUS DISEASE AND
PERSONAL IDENTITY

Indigestion had been bothering a young Bostonian for more than a year by the time he sought help at the office at 190 Marlborough Street in June 1902. He had also been suffering from nervous episodes, during which he would "do a lot on excitement," or so Cabot learned when he interviewed him. Examining him closely and detecting no other obvious troubles or abnormalities, Cabot entered his diagnosis, debility, in the young man's chart on their first visit and offered some advice about a digestive medication, outdoor sleeping, and arrangements to track his weight. The young man's problems persisted, however. He returned to the office in 1906, 1907, 1909, and again in 1910. Some features of these visits remained the same. He continued to report a variety of physical, psychic, and emotional discomforts, while the examinations and testing continued to turn up no evidence of any specifiable physical disorder. So in this sense the young man's predicament was stable. Yet one thing did change markedly over the years—the diagnosis. The initial diagnosis in 1902, "debility," changed briefly to "psychasthenia" in 1906, then to "neurasthenia" in 1907, and finally to "psychoneurosis" in 1909.[1] All fell within the family of general nervous disorders as they were understood at the time, manifesting a mixture of physical, emotional, and psychic distresses without the evidence of physical abnormality.[2] But this changing diagnosis was all along moving in step with a larger, general trend in Cabot's practice, as can be seen in table 5.1. Cabot's favored term for nervous disease changed over this same period from "debility" to "neurasthenia" and then to "psychoneurosis." This pattern of change in diagnosis, exemplified so neatly in the young man's case, reflected a gradual transformation in the management of nervousness for Cabot and many of his peers in the early decades of the twentieth century.

The general nervous diseases, as Sir William Osler classified them in his famous medical textbook in 1892, were conditions defined primarily by the impairment of mental and emotional function. This had been the definition throughout the nineteenth century, and so it remained for Cabot. But the physical characteristics of these conditions had by the early twentieth century become increasingly incidental. The bodily symptoms of nervousness, their acknowledgement and their treatment, lost their claim on the physician. Historians have noted this fundamental shift in the handling of nervousness in late nineteenth-century medicine, particularly in relation to the conditions of hysteria and neurasthenia, and have speculated about the reasons for the change and its effects in practice. Cabot's office records provide an opportunity to examine the details of daily negotiations with patients suffering from nervous disorders in the context of Cabot's own tidy recapitulation of this gradual change. Nervousness became less a bodily condition and more exclusively a property of the emotions and thought, with critical implications for the expression of medical authority in daily practice.[3]

This apparent restlessness in Cabot's diagnostic terminology bespoke a fundamental tension that he faced in managing nervous disease. The medical techniques and rationales that were available in his clinic promised to resolve basic questions about physicians' power and responsibility to handle the troubles of their patients. The physical nature of disease was made persuasively evident through x-rays, blood tests, and biopsies. Thus exposed, disease was liable to treatment with procedures and medications that had visible, measurable effects. This rigorous schema at the center of practice yielded a basis for medical authority that seemed clear, impersonal, and verifiable. In practice, however, this schema functioned as much by exclusion as by application. Nervous disorders marked an area at the boundaries of natural disease where the power of these methods tapered off. Nervousness as a medical condition seemed to lie halfway between the real and the imagined, its hidden sources not easily found. In some cases, a nervous patient was meant to accept proof of the absence of a physical disease as the principal medical evidence for nervousness. Cabot offered the diagnosis of "syphilophobia," for example, when a patient's fears about the symptoms of syphilis could be laid to rest through diagnos-

TABLE 5.I   THE DIAGNOSIS OF NERVOUS DISORDERS BY RICHARD CABOT, 1900–1915
(in percentage)

	Debility	Neurasthenia	Psychoneurosis	Not sick
1900	12	5	0	0
1901	12	2	0	0
1902	7	4	2	0
1903	10	10	2	0
1904	11	6	0	0
1905	3	10	0	0
1906	2	10	1	1
1907	4	10	4	1
1908	1	11	5	0
1909	2	2	16	2
1910	5	1	21	5
1911	1	1	13	4
1912	0	1	9	6
1913	1	0	12	6
1914	1	1	13	6
1915	1	0	17	7

Note: Each patient whose case was recorded for the first time in a given volume of records had a diagnosis set down in the index to that volume. The figures, which have been rounded off (e.g., <0.5% is given as 0), show the major nervous disorders and those diagnosed as not sick as percentages of all diagnoses each year (e.g., 12% of all diagnoses in 1900 were for debility).

tic testing—a Wassermann serum test in this instance. Nervousness generated by fear of the disease was diagnosed by proof that the disease was absent. The persuasiveness of such negative evidence was, however, limited. Even as Cabot's patients acknowledged the potential power of diagnostic technique to exclude conditions, they often rejected its application to their own case. The objective, impersonal sources of medical power that Cabot vigorously defended seemed limited in their ability to define the conditions of nervousness persuasively.

The gradual transformation in nervous disease in the clinic in part reflected Cabot's efforts to sharpen and clarify these difficult boundaries of medical authority. But this transformation also mirrored wider changes in theories of nervousness. The young Bostonian mentioned earlier might

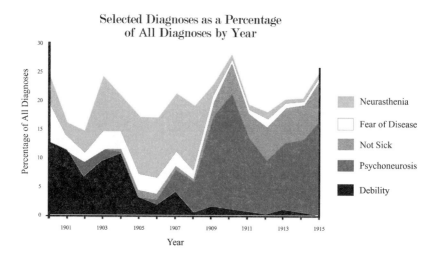

Selected Diagnoses as a Percentage
of All Diagnoses by Year

Neurasthenia

Fear of Disease

Not Sick

Psychoneurosis

Debility

Nervous disorders diagnosed by Richard Cabot, 1900–1915. A graphic depiction of the data in table 5.1, showing nervous disorders as a percentage of all diagnoses.

reasonably have complained that his changing diagnosis was subject to the whims of medical fashion, with popular diagnoses replacing one another in turn. Someone familiar with contemporary medicine would have discerned an interesting pattern in Cabot's evolving system of diagnosis for nervousness. The change from debility to psychasthenia in 1906 coincided with a visit to Boston by the eminent French physician Pierre Janet, who coined the use of this term for a nervous condition. Professor Janet presented several lectures at Harvard Medical School that same year, which epitomized growing efforts among European physicians to shape psychodynamic models for nervous disease, and which received special attention among Cabot's colleagues in Boston.[4] Psychasthenia was a disease of the times in 1906.

Neurasthenia, the young Bostonian's next diagnosis, was similarly the object of considerable medical and popular attention at the turn of the century, so that the emergence of this diagnosis to replace an older term like "debility" seems equally influenced by popular trends. Cabot's next change from neurasthenia to psychoneurosis in 1909 coincided in a similarly fortuitous way with a celebrated series of lectures at Clark University near Boston given by Sigmund Freud.[5] By this time, neurasthenia was beginning to lose some of its cachet, as American physicians moved away from a formal, medically specific use of the term. Especially in Boston, a growing interest in the psychodynamic origins of nervous disease was coalescing around the investigation of the condition of psychoneurosis by Freud and his American colleagues.[6]

But if Cabot's changing diagnostic nomenclature was following the fashions of the times, it also marked a substantive transformation in his management of nervous disorders, as again neatly exemplified by the gradual change in the same Bostonian man's care. Debility, his diagnosis in 1902, was a well-established nineteenth-century term for an undifferentiated condition of bodily, emotional, and mental exhaustion. The young man suffered from distressing episodes of intense mental "excitement," along with indigestion, so that Cabot's advice in 1902 about fresh air, digestive medications, and body weight showed careful attention to the patient's general physical well-being and physical symptoms, as well as his mental and emotional health. By 1907, neurasthenia had replaced debility as the diagnosis, implying much shallower claims about the young man's bodily condition and care. Although neurasthenia was originally characterized in the nineteenth century as a disorder in the physical energy of the nervous system—the "nerve force"—by the early twentieth century, the term had come to indicate a depletion of the emotional and mental energies instead.[7] Nervous disorders like neurasthenia and hysteria, which had long drawn the attention of physicians for both their mental and physical effects, were gradually losing their claim to describe significant physical alterations in the body. The shift in Cabot's practice away from inclusively physical models of nervous disease was complete by 1909 when this young man was diagnosed with psychoneurosis, a disorder limited rather strictly to the mind. His nervousness was now comprised of emotional and psychic disturbances, readily distinguishable from the electrical energies that pulsed through the spinal cord and nerve fibers. Correspondingly, the young man was noted in his medical chart in 1909 under the diagnosis of psychoneurosis to be "too much influenced by mother," a problem that "lingered on in subconscious form." Although Cabot still noted his patient's physical symptoms in his chart, he reported that this young man now accepted that "his trouble has been moral + mental," a conclusion apparently acceptable to the patient himself, because he took practical steps to address this aspect of his condition. Following his 1909 visit with Cabot, he went to the pioneering American psychiatrist Adolph Meyer in Baltimore for psychoanalysis.[8] Fresh air and improved digestion no longer seemed to be major concerns for the physicians who were treating his nerves.

For Cabot and many of his peers, the evidence of nervousness seemed increasingly to lie in the details of a patient's personal conditions and social circumstances, in experiences that "lingered on in subconscious form." A fine separation emerged in practice, like a line of perforation where the sources of the physician's influence might be pulled gently apart. Technical control of disease differed from influence over patients. Knowledge about the particular circumstances of an individual patient had long been a source of persuasive authority for physicians, who might claim to diagnose and treat a patient based on a detailed familiarity with his or her life and circumstances. Yet for Cabot the relevance of the personal and social features of a medical case might matter most in nervousness, an area of practice where the legitimate powers of technical medicine seemed least applicable. It was a separation that sharpened the clarity of his control over physical disease but equally served to make the personal and social features of illnesses as contentious as nervous conditions.

Cabot's sharpening of the separation between the nervous and the physical diseases heightened tensions in the management of nervous patients. People diagnosed with a nervous disorder like debility, neurasthenia, or psychoneurosis typically suffered from troubling physical symptoms. Yet the symptoms of nervousness were gradually losing a medically endorsed connection to physical, bodily care, as the qualities of nervousness were pushed aside in medical scrutiny of the physical body. The young Bostonian seemed to accept a shift in focus away from his bodily symptoms like indigestion, but others among Cabot's patients passionately resisted the notion that their troubles were all in their heads. With the diagnosis of debility, and the more somatic versions of neurasthenia, Cabot and his peers had held out the possibility that physical, bodily treatments and explanations applied to nervous conditions. The nerves had long been seen medically as mediating between personal circumstances and physical well-being. Nervous diseases held an important middle ground in late nineteenth-century medicine, linking stressful circumstances and social conditions with physical breakdown, making "shattered nerves" a service-able category of clinical practice.[9] By 1909, when Cabot took up the diagnosis of psychoneurosis, he was limiting his attention in nervous cases more strictly to the personality and mind. A persuasively eclectic model of nervousness no longer seemed available.

Cabot attempted to use the leverage afforded by a technical knowledge of disease to organize a clinical response to nervousness. But his efforts did more to highlight the difficulty than to dispel it. One possible avenue of effect would be to use expertise in physical disease to define and complement a separate expertise in managing disorders of the personality. Cabot often parceled out the care of his nervous patients to psychoanalysts or spiritual counselors and other lay mental healers. In his most structured venture of this kind, he entered briefly between 1906 and 1909 into an agreement with Elwood Worcester and the spiritual advisors of Boston's controversial Emmanuel Movement to send them ailing nervous patients whom he had certified as free of bodily disease, as described briefly above in chapter 3.[10] In his private clinic, efforts to convince patients of a psychic source for nervousness through diagnostic testing met with comparable difficulties, leading to the manipulation of technique for merely persuasive ends. In one early instance, he even experimented with and rejected what looks like placebo diagnosis, in which the diagnostic process served only to create the appearance of attention to physical disease. These unsuccessful trials left him with the need to muster other resources for medical control over nervousness. His handling of nervousness came to draw instead from longer-established sources of authority in the social hierarchies of race, class status, and gender.

Debility: Nervousness in the Mind and in the Body

Roughly one-fifth of all the new patients seen in Cabot's office during his busiest years, between 1900 and 1915, were diagnosed as having one of the general nervous disorders (table 5.1). In this period, Cabot drew from a wide range of diagnostic terms to characterize the condition of nervous patients, including grand hysteria, apprehension, syphilophobia, sexual neurosis, and nervous exhaustion.[11] Such nervous overexcitements and depletions shaded off into a smaller set of idiosyncratic diagnostic terms, inspired perhaps by a particular patient, so that Cabot would on rare occasions diagnose someone as suffering simply from overwork, loneliness, or high living. This practice of designating personal behavior or a circumstance of life as a disease category had a minor variant in the designation of aberrant habits as diagnoses. For instance, Cabot also, albeit rarely, diagnosed conditions like excessive indulgence in tobacco, poor hygiene, overeating, morphinism, or alcoholism, keeping more commonly to the typical practice of diagnosing the pathological effects of these aberrant habits. A comparable set of terms captured the harsh effects of the circumstances of the patient's life, leading to the diagnosis in a handful of cases of unemployment, starvation, or destitution. Cabot seldom used such diagnoses, but they illustrate the wide range of terms that he could drew on to characterize the problems of his patients. Within this eclectic mix of diagnostic terms, however, Cabot's three major categories of general nervousness stand out quite plainly: debility, neurasthenia, and psychoneurosis, each dominating his diagnostic practices for several years in turn. Debility came first.

Debility held a favored place in Cabot's practical diagnostic system from the first days of his practice. Its role as one of the general nervous conditions was evident from its relationship to other concurrent diagnoses within his practical nosology. Debility occupied an important middle position between clear-cut physical abnormalities and profound mental disturbances, like insanity—the same place to be held later by neurasthenia and psychoneurosis. It was a stable classificatory niche that accounted for a large variety of bodily symptoms in patients, in addition to emotional and psychic difficulties and discomforts. In sorting through patients' problems in the office year after year, Cabot diagnosed roughly 65 to 75 percent as afflicted by some specific bodily disorder, such as nephritis, colitis, bronchitis, peritonitis, gastritis, cellulitis, angina, gallstones, peptic ulcer, anemia, leukemia, various cancers, diabetes, Graves' disease, hypertension, malaria, tuberculosis, and syphilis, among many others. He diagnosed a much smaller number of people with profound mental disorders, identifying such conditions as manic depressive insanity, dementia praecox (later called schizophrenia), melancholia, or senile dementia. It was rare for anyone to come to this clinic for the first time feeling well, so the remainder of the patients with neither markedly disordered thinking nor obvious bodily abnormality had something else wrong with them. Cabot gave almost every patient a diagnosis, and general nervousness made the largest single contribution to his classifications.

At a peak in the years 1900 and 1901, roughly 12 percent of all new patients in Cabot's clinic received the diagnosis of debility (table 5.1). It was a term widely used by late nineteenth-century physicians in Boston to categorize people who felt physically ill but whose sickness was not easily isolated or defined. At the busy, prominent Massachusetts General Hospital, for example, in the 1880s and 1890s, doctors commonly applied the diagnosis of general debility to patients on the medical wards. Among twenty-five patient-cases selected using random sampling from one volume of the hospital's medical cases in 1894, three patients (coincidentally, 12 percent) had this diagnosis, "general debility," on their records. Other patient records from the hospital in the 1880s and 1890s show similar diagnostic practices throughout the period.[12] The diagnosis of debility, while vague about the source of the patient's troubles, seemed not to exclude the possibility of associated bodily disorders. One patient diagnosed at MGH with debility, for example, exhibited recurrent fevers. Almost all these patients with debility had significant bodily symptoms that had brought them to the hospital, since during these years, obvious mental illness was typically not accepted as a reason for admission. A diagnosis of debility in this setting carried little specific information about the bodily problems of these patients, but allowed that there was physical significance in their frequent complaints of weakness and exhaustion.

Cabot used debility in a similar way in his office to indicate an undifferentiated condition that included physical, emotional, and mental exhaustion. Debility was his final diagnosis, for example, in the case of a middle-aged woman from the nearby town of Somerville, Massachusetts, who in 1901 described her problems with headaches, constipation, and weakness to him. She had to contend, her chart recorded, with "a troublesome set of boys over 15"—evidently worthy of note as significantly related to her difficulties. A physical examination and laboratory testing of her blood and urine revealed no abnormalities. The combination of her symptoms, her circumstances, and the absence of an identifiable physical aberration presumably led to the diagnosis "debility." Just as in MGH, debility in Cabot's office did not require the absence of major physical symptoms, conditions, or signs to be applicable. A twenty-five-year-old married man involved in "hotel work" came in May 1900 reporting to Cabot a series of ear infections and a problem with insomnia, duly noted in his chart. He too received the diagnosis of debility. But debility did not apply to people whose symptoms were accounted for with the evidence of a specific, localized bodily abnormality. A woman who reported a year of "spells" in her stomach without other manifestations of stomach disease received the diagnosis of debility. In contrast, another young woman who reported similar episodes of stomach discomfort but added that she had "seen some blood," presumed to come from her stomach, warranted the diagnosis of ulcer dyspepsia, rather than debility.[13] The identification of an ulcer located her medical problem in her stomach, while the other woman's debility presumed some overall disturbance of the mental and physical constitution. Debility was treated as a general disorder that was compatible with distinct nervous and bodily problems but not reducible to a particular bodily site.

Debility was a serviceable diagnosis in Cabot's daily medical practice. It acknowledged the physical implications of otherwise vague weaknesses, spells, and disturbances of function. Cabot was explicit about the mixed causal elements that this diagnosis accommodated. Commenting on a case diagnosed with debility from 1899, he wrote that two things were clear: "first that psychic causes enter into it, and second that they are not the whole of it."[14] Debility provided a flexible and persuasive explanation for a patient's symptoms by permitting consideration of a variety of causes. In addition, debility supported the eclectic use of physical remedies for relief. Cabot offered many different treatments for the problem of debility, including restorative tonics and symptomatic remedies. Another young woman, who reportedly had "profuse nerves," received the diagnosis of debility in April 1900 from Cabot. Like the woman with the troublesome set of young boys under her care, she also had circumstances in her life that seemed directly relevant to her condition. Cabot noted tersely that she had recently been affected by "losing a sister and having a baby"—an odd juxtaposition, perhaps prompted by a comparable strength in their life-altering effect. Since this woman was also "pale," her treatment consisted of Blaud's pills, a proprietary tonic used for "building up" the blood.[15] Debility invited the therapeutic application of both medicines and general hygienic measures with alterations in daily habits and routines as an expected part of the prescription. Patients who suffered from debility around 1900 received a variety of different medications from Cabot, such as reduced iron, hypophosphites, cascara, potassium iodide, bismuth subnitrate, and beta naphthol.[16] Yet this comfortably eclectic mix of therapeutic approaches did not last out the first decade of Cabot's practice.

Whatever its usefulness in practice, debility soon lost favor both with Cabot and among his prominent Boston colleagues. By 1900, physicians at Massachusetts General Hospital had dropped the diagnosis almost entirely. Two hundred forty-five randomly selected cases from seventy-one volumes of medical records (roughly 7,000 cases) from the East Medical Service at MGH between 1900 and 1910 include no single occasion of the use of debility as a primary diagnosis.[17] In Cabot's practice, too, by 1905, debility had fallen from the high of 12 percent to 3 percent of all new diagnoses, and by 1908 to 1 percent (table 5.1). Cabot replaced the "debility" with "neurasthenia" as his favored term to describe general nervousness.

The sources for this change in diagnostic habits were doubtless complex. In Cabot's case, a growing devotion to the project of exact diagnosis was one potential factor. By 1910, his Clinicopathological Conferences at Massachusetts General, described earlier in chapter 3, were in full swing, and whatever the diagnosis revealed at the conferences by autopsy, it could never be simply "debility." There was no pathological lesion of debility. The changes in diagnostic protocol suggested here reflected larger trends in the use of diagnostic testing by Cabot, by colleagues in the hospital, and to an extent by physicians nationally. The convincing—if still not readily sustainable—power of technical evaluation to reveal hidden diseases was growing. The wider application of x-rays, serology, blood counts, surgical biopsy, and chemical analysis of body fluids that provided the evidence for the Clinicopathological Conferences also acted as a reminder and confirmation of this diagnostic power. Diagnosis would name the specific physical disturbance that manifested itself as a general weakness or a vague debilitation. Perhaps there was a hidden cancer, syphilis, or a remediable anemia. New diagnostic technologies made it possible to find these problems. So when the diagnostic process revealed no aberrations, the question inevitably arose of whether any bodily abnormality remained behind associated with debility. Under the gaze of more confident diagnostic analysis, debility followed a path seen with other nineteenth-century diagnoses as it was redefined as a general symptom of a disease, rather than a disease in itself.[18]

The term "debility" remained valuable for describing a patient's overall medical condition, even as it disappeared as a diagnosis for a general nervous disorder. In Cabot's practice, it lingered on as a term modifying other more specific afflictions. It preserved a sense of the comprehensive exhaustion and dysfunction that resulted from other disorders. Cabot used debility only rarely as a primary diagnosis in his medical office after about 1906. An occasional patient did, however, receive a diagnosis such as "debility (domestic abuse)," "debility post miscarriage," "post-typhoid debility," and even "debility of adolescence," as one girl's troubles were characterized.[19] "Debility" neatly summed up the wide-ranging influence on a person of various events, afflictions, and diseases. At Massachusetts General Hospital too, the term "debility" continued to find use as a modifying characteristic of other medical disorders or medically significant events, as with the woman who carried the set of diagnoses in her hospital chart of "mitral regurgitation/debility/cardiac asthma."[20]

Therapeutic trends may also have contributed to this declining use of the diagnosis of debility. The movement for physiological therapeutics, described above in chapter 4, emphasized the value of medical treatments that had specific bodily effects that could be measured and tracked. Debility, however, evinced no such traceable abnormalities. Was it fair to use a medication like "reduced iron" in debility, if the blood counts were not a matter of concern? Parallel trends in the application of specific mental therapeutics carried similar implications. A growing literature and clinical interest, especially in Boston, about the expert management of the nervous disorders through suggestion, hypnosis, psychoanalysis, and other mental interventions may have encouraged a more precise separation between the nervous and the bodily disorders. By the turn of the century, physicians in Boston were turning their attention to defined psychological methods of treatment. In the first decades of the century, established psychodynamic theories gained renewed support as they were allied with and transformed through the specialized doctrines of psychoanalysis.[21] Pierre Janet's and Sigmund Freud's visits to New England in 1906 and 1909 respectively were markers of a widening interest in focused, instrumental uses of mental therapeutics in medicine.

Cabot's concern about the boundaries of legitimate medical influence over patients also likely figured into this change in diagnostic practice. His critique of placebo treatment was directly implicated in this problem of properly restraining the physician's response to nervous patients. By treating debility with reduced iron or naphthol, a physician risked suggesting that specific physical disturbances needed to be corrected—a suggestion easily exposed as misleading. Cabot drew attention to this dilemma in the management of nervous disorders in a critique of the methods of the prominent American neurologist Dr. S. Weir Mitchell that he published in 1896. Neurasthenia, Mitchell had taught, could be treated through enforced rest and overfeeding of the patient, intended to build up depleted stores of nervous energy through the promotion of "blood and fat." Cabot examined the physiological basis for this treatment with disappointing results. He conducted a series of blood counts on patients suffering from untreated neurasthenia and found them to be normal. He challenged the notion that Mitchell's treatment to increase blood and fat would in itself make any appreciable difference "when the patient has already plenty of each." Any beneficial effects of the treatments might be obtained as easily, Cabot argued, through "the same *régime* as a matter of suggestion."[22] In other words, rest and feeding might act beneficially on the psyche of the patient, but not on the blood. But did such covert psychic influence represent a misuse of the physician's authority?

The force of this dilemma was made clear to Cabot in applying precise technical methods to his daily private practice. The middle-aged woman with the troublesome set of boys who came to see him in 1901 received a diagnosis of debility. Cabot treated her with Blaud's pills, a medication that he used also in the treatment of anemia, to improve the level of red cells in the blood. But this woman's blood counts, like those of Mitchell's neurasthenics, were normal. Following his entry on the prescribed treatment, "Rx: Blaud's," Cabot dutifully added the notation "(placebo)." The pills actually seemed to help somewhat in relieving this woman's symptoms, as Cabot noted in his chart.[23] But he had already questioned treatments like Mitchell's to improve the levels of blood cells in patients with seemingly normal counts. By appending the word "placebo," Cabot forced a sharp distinction between the targeted treatment of a disease and the manipulation of the patient's expectations and perceptions. This was a distinction that was difficult to retract, once made. Specific medical treatments for debility might qualify as nothing more than placebos, whose use in medical practice Cabot would soon feel compelled to renounce.

When he abandoned debility as a favored nervous diagnosis, Cabot simultaneously altered his strategies for treating the nervous disorders, preserving consistency with a newer model of nervousness. His use of specific drugs to treat nervous diseases declined, while his use of psychological methods and justifications became more pronounced. His therapeutic response to nervous disease shifted to follow the movement of nervousness out of the body and back exclusively into the mind. The shift was subtle; and the pressures of ordinary medical practice surely acted as a corrective to any single, dogmatic approach to treatment.[24] Certain elements of care for nervous disease remained constant. Directions for personal habits, hygiene, and daily regimen continued as a routine part of the advice that he delivered about the care of the nerves. Such measures for debility, and to a seemingly lesser extent for neurasthenia, still linked up with specific drug treatments. Later in the treatment of psychoneurosis, however, as drug prescriptions became rare, the management of personal habits and hygiene became linked instead to psychodynamic and "talk" therapies.[25] By 1910, Cabot had largely dropped debility and neurasthenia as diagnoses, while in the same year diagnosing roughly 20 percent of all new patients to his office with psychoneurosis. In addition to advice on outdoor activities, diet, and occupation, patients diagnosed with psychoneurosis received treatments that Cabot described variously in the charts as talk, advice, general philosophy, or "autosuggestion + explanation."[26]

Advice about personal habits and hygiene, while it remained a stable part of therapeutics for nervousness, bore different connotations during the different phases of Cabot's evolving practice. As the eclectic bodily and psychic model of debility gave way to the purely psychological model of psychoneurosis, the accompanying advice about personal hygiene shifted subtly in its intent and significance. In the treatment of psychoneurosis, Cabot's advice about personal habits and hygiene linked up easily with efforts to train the patient's sentiments and psyche. In the treatment of debility, in contrast, such advice connected with efforts to support the patient's physical health and meshed neatly with prescriptions for medications to manage physical symptoms. On two successive days in May 1900, for example, two women came to the clinic with nervous disorders, one married and in her fifties and the other single and almost out of her twenties. Both were diagnosed with debility. Their charts noted significant physical difficulties as well as nervousness: the older woman had a cough, had lost weight and had recently stopped menstruating, while the younger woman was "thinner for two years . . . [and] tired," although "not at all pale" on examination. Examination and testing, however, yielded no further physical source for their various troubles. A common hygienic agenda seemed to motivate the advice to both women to get outdoors for fresh air, and for the older woman to keep her windows open at home. Cabot might have had in mind a common wisdom of the day for staving off incipient tuberculosis. But he had not diagnosed that condition and was in fact offering something more. He prescribed several medications, cascara and heroin for the older patient, and Blaud's pills, nux vomica, and gentian for the younger.[27] The use of these specific drugs, in connection with the emphasis on physical symptoms, made the advice to get more outdoor air seem part of a set of general recommendations about physical and environmental influences on bodily health. Cabot made no evident connection between fresh air and lifting the spirits, for example.

In contrast, in his later treatment of psychoneurosis, Cabot linked very similar hygienic advice about personal habits and regimen to the nurturing and discipline of the psyche. A common recommendation for the treatment of psychoneurosis was "Rx: work."[28] When taken in combination with "Rx: talk," this exhortation connected the discipline of daily routine to mental discipline, or at least to a distraction from mental anxiety or confusion in the simplicity of a daily routine. Advice about fresh air seemed in this context more about mental and emotional vigor than about alteration of the patient's physical condition. A single man in his twenties whose chart in 1912 carried the diagnosis of psychoneurosis was, for example, deeply concerned about the condition, writing to ask Cabot "whether you think that my *nervousness* may possibly be my ruin?" He visited and wrote Cabot often over a period of several years. His treatments at one stage included "a morning walk of about four miles at a moderate pace, followed by a bath & short rest." He exercised regularly out-of-doors and took up residence in a rural setting for a short period as an extension of his treatments. His chart bears no evidence of specific drug treatments. Instead, rigorous daily hygiene was combined with psychoanalysis. Cabot recorded that the New York analyst J. Ramsay Hunt had "turned him inside out & says whole trouble is sexual beginning very early & suppressed."[29] In the context of psychodynamic treatment, the discipline of a daily walk in the fresh air seemed more relevant to the state of the patient's mind than his body.

Cabot's medical advice about personal habits and hygiene preserved a useful flexibility in its application to the nervous diseases. Even as his working model for nervousness shifted away from physical debility toward psychoneurosis, he continued to offer advice about specific hygienic routines. Writing to a woman diagnosed with neurasthenia, he suggested as an initial treatment that she seek "routine and tangible usefulness in some field, the holding down of a job which won't go on unless you are there and which give[s] you the physical as well as mental stimulus of being of definite use." Here was Cabot's "Rx: work" applied to the treatment of nervousness. His letter did not deny the beneficial physical effects of a disciplined personal routine but merely offered a pragmatic consideration that in cases of psychoneurosis such as hers, "the mental element is the most available one for treatment; that is to say we know no drug that will help you."[30] In his treatment of nervous diseases, Cabot drew back from a claim of beneficial control over bodily functions through medication. Personal habits and environment nonetheless remained valuable avenues by which therapy might influence the important psychic elements of nervous disease.

Neurasthenia and the Contentiousness of Nervous Disease

As debility disappeared from Cabot's practical diagnostic schema after 1905, neurasthenia assumed its place, accounting for as much as 10 and 11 percent of all of his new diagnoses in 1905 and 1908 respectively. Cabot's interest in this condition belonged to a well-established tradition within American medicine dating back to the nineteenth century but equally reflected a general early twentieth-century preoccupation with a condition widely discussed and debated in the popular newspapers and weeklies. Physicians by the twentieth century had embroiled neurasthenia in considerable debate and controversy over its causes, treatment, and even existence, perhaps in response to its very success in the public's eye. It remained, nonetheless, a widely diagnosed medical condition around the turn of the century. Originally conceived of and named in the 1880s by the American physician George Beard, neurasthenia attracted wide attention in the New England of Cabot's day.[31] Even among patients sick enough to enter a general hospital at the time, neurasthenia was a frequently identified problem. It was the most common diagnosis applied in the year 1898 at the New England Baptist Hospital, tied with influenza for that honor during this last year in which the hospital published its diagnostic statistics.[32] Roughly 9 percent of more than four hundred patients diagnosed on the medical wards of St. Vincent's Hospital in Worcester, Massachusetts, in 1908 were found to have neurasthenia on discharge.[33] Neurasthenia was a similarly common diagnosis on the wards at MGH during the early period of Cabot's training and work.

Although neurasthenia claimed a substantial place in the medical literature of the early twentieth century, it was coming under increasing criticism in its clinical use. Did it qualify as a disease or merely as a placeholder for other more exactly defined and perhaps more critical diseases? Physicians claimed that in some instances, the diagnosis of neurasthenia resolved upon careful scrutiny into any of a number of important physical disorders. In the pages of the *Illinois Medical Journal,* one physician described how a diagnosis of neurasthenia provided only a distraction from more important underlying pathologies. Under examination by an astute diagnostician, a patient presumed to have neurasthenia often turned out to be suffering instead from "tuberculosis, dysthyroidea, dental disease, genito-urinary inflammation, [or] syphilis," for example. The challenge for the diagnosing physician was to reveal the presence of such "organic disease" and to avoid being fooled by "psychic counterfeits" like neurasthenia.[34] Physicians had to attend carefully to disorders that in some cases lay in the mind and the imagination of the patient and in others were convincingly detectable in the body. By diagnosing neurasthenia too readily, physicians risked overlooking crucial bodily afflictions.

While some physicians, like this Illinois author, admonished their colleagues not to miss the actual physical abnormalities that hid behind a diagnosis of neurasthenia, others claimed instead that neurasthenia was simply an imprecise designation for, or perhaps a mild *forme fruste* of, mental disease. Physicians in the specialty of neurology, although they were initial claimants to expertise over neurasthenia, became in the early twentieth century especially vocal critics of the diagnosis, who disputed its value in clinical practice. Neurasthenia in 1900 still carried with it the vestiges of its nineteenth-century characterization as a physical disorder of the nerve force. Yet neurologists noted that patients with neurasthenia typically lacked the visible, measurable defects in the central and peripheral nervous system that their specialized technical skills would permit them to detect. Neurasthenia and its companion disease hysteria were in the opinion of one neurologist simply "the milder forms of disordered mental states" that "commonly simulate physical and organic disease."[35] It may have been particularly irksome to a specialist in the diseases of the nervous system to encounter the diagnosis of neurasthenia, carrying the lingering implications of a well-defined abnormality of the nerves. The prominent neurologist George Lincoln Walton registered this complaint, pointing out with disdain that his colleagues on the medical service at MGH still diagnosed neurasthenia routinely in 1909. He and other neurological specialists who had confident knowledge of the nervous system understood that these patients with neurasthenia were "cases without known organic basis," that is, afflicted exclusively with mild mental disturbances. He enjoined other physicians to follow the neurologists' lead and "just call them all psychoneurosis and leave it at that."[36] The use of the term "psychoneurosis" did away with any residual suggestion that there was a physical abnormality in the nerves underlying the patient's condition. Cabot's changing management of general nervous disorders was consistent with this usage.

Although appealing to a neurologist like Walton, the idea that neurasthenia could be rolled back out of the body was a contentious one for many of the patients who visited Cabot's clinic. An illness completely unrelated to physical derangements might seem less serious, less concrete, or even entirely unconvincing. People writing to Cabot pointed out several shortfalls with his changing diagnostic practices. First, they worried that accepting nervous disorders as purely psychic might lead physicians to miss other threatening bodily diseases. This concern of patients echoed that of the physicians who warned that neurasthenia cloaked important undiagnosed problems. One Boston woman writing in 1909 recalled to Cabot in a fury an instance in which another physician of hers had belittled her breathlessness, "laughing and calling it 'only nervousness.' " Later, she developed a severe heart disorder, an initial stage of which she believed had earlier caused her breathlessness.[37] The physician's attribution of her condition to nerves seemed to her a dangerous neglect of a significant bodily problem. In addition, she identified a dismissive attitude toward her laughingly diagnosed nervousness. The pronouncement that the trouble was only nerves seemed patronizing to her. Cabot himself observed in regard to an individual diagnosed with neurasthenia whose symptoms were later explained by a gallstone that such a diagnosis was "often adding insult to injury."[38] Although he did not distinguish the insult from the injury explicitly, he implied that the diagnosis of neurasthenia was itself insulting, as well as evidence of an injurious neglect of physical disease.

People who came to Cabot's clinic also expressed the related concern that nervous conditions were disturbing conditions in themselves. Having received a diagnosis of neurasthenia, a woman in her twenties who came from Connecticut for a consultation expressed a deep dismay about the condition when she wrote back following her visit in 1912. Cabot's initial concern at the visit—apparently shared by her—was that she had tuberculosis. He followed up on his interview and examination by sending her for additional testing. After this careful diagnostic evaluation failed to demonstrate the presence of tuberculosis, Cabot wrote back reporting this result and advising her about a further course of care. In his letter, he described her condition as "a congenital weakness of the nervous system, what we ordinarily call neurasthenia." Summarizing the normal results of testing for tuberculosis in his letter, he attempted to dismiss any of her residual concerns about the disease.

His patient met this news with a strong expression of her ambivalence. "From your note I gather you seem to feel most of my trouble is mental," she wrote back. "This diagnosis I suppose is correct—I should be inclined to believe it so for the simple reason you made it. It ought to comfort me I suppose, but I find it fills me with a disconcerting hopelessness. A mental trouble is such an intangible thing."[39] Neurasthenia would seem to have been a less harmful condition than tuberculosis, then frequently fatal. So why did she not accept the diagnosis of nervousness as comparatively good news? Perhaps the diagnosis of nervousness suggested to her the specter of more severe mental disease or seemed to hold out less chance of recovery. She did not say.

What she did say about nervousness is quite remarkable. Neurasthenia did not provide the same level of certainty about or justification for her illness. It was intangible and represented little explanation for the debilities that she experienced. She went on in her letter to explain that if she had actually had tuberculosis "I would forgive myself for utter exhaustion after a walk of a mile or even find it in my heart to excuse the mental depression which as invariably comes when I need to take a pill!"[40] Even depression, paradoxically, seemed more easily excused when it could be ascribed to a physical disorder like tuberculosis. A dreaded disease like tuberculosis, she suggested, was at least concrete and substantive, providing a solid explanation and justification for feeling sick. Nervous disorders like neurasthenia that were being cut loose from their bodily moorings could not so easily serve such ends, at least in the context of a discussion with one's physician.

This sharp demarcation between the physical and the mental was not always easy to maintain in accounting for the complex problems of personal health. Cabot sometimes expressed his own dualism as a hypothetical distinction that, while valuable, was difficult to act upon in practice. In the treatment of physical diseases, for example, he sometimes acknowledged the inseparability of somatic and psychic health as an awkward reality. Writing to a sympathetic medical colleague in 1916 about the treatment of a patient diagnosed with the degenerative neurological disease of paralysis agitans (what might later be called Parkinson's Disease), Cabot ended up by recommending what he admitted was the less effective of two possible treatments for the condition. The treatment that he rejected would be too liable to depress the patient, as he explained it, "although this might be the best [treatment] were there no need of considering the patient's mental condition."[41] The patient's mental condition, although of seemingly secondary concern, still had a bearing on decisions about treatment, Cabot conceded. The more physically effective treatment was "best," as he deemed it, but not practicably separable from its depressive side effects.

Patients writing to Cabot tended on occasion to blend together the concepts of nerves as mental and as physical entities. The stressful circumstances that affected their emotional temperament also caused their specific physical reactions. One man writing to Cabot immediately after his return home from Massachusetts General Hospital noted that he was continuing to suffer from "pain and temperature [and] general discomfort." He included a small handwritten chart of his body temperature as registered at home to demonstrate the details of his fevers, running as high as 102 degrees. He reported that he was getting better, and speculated about the possible sources of his fever and pain. "I wonder," he concluded his note, "if it was not caused by nerves at [the] hospital and by the very rough r[oad?] home?"[42] His condition was improving: the nerves, the fever, and the pain together. The connection between nerves and physical health seemed a commonsense explanation to him. The similarity and overlap of emotional and physical states seemed just as natural too for another man who wrote that "any worry or excitement seems to make me feel I still have those ulcers."[43] Worries might exist primarily in the emotions, but it felt as though they were intimately connected with the problems that arose in the stomach.

Even when Cabot's correspondents acknowledged the theoretical difference between nervousness as either physical or mental, they often had concerns that overwhelmed the usefulness that they found in this distinction. A woman in her seventies from nearby Brookline had more pressing matters than Cartesian dualism to consider when she wrote. She was sick. Following a visit to the office, she penned a letter to let Cabot know that she had in fact been "slowly improving in general conditions." She explained that her improvement related to a restoration of the nerves, but that several influences were likely at work. She entertained the question of what different factors were involved. Her recovery, she speculated, "has been due to the gradual nourishment of the nerves brought about by rest and proper food." However, determining if this improvement were not due "more to mental healing ideas," she confessed, "would be difficult [to] know." Therapy for the nerves required the healing of the mind as well as the nourishment of the body. Her conclusion was that "undoubtedly each has had a share in bringing the result."[44] The ambiguity in the meaning of "nerves" made a strict choice between bodily and mental healing unnecessary. Surely what counted was the result. To secure reliable care, the practical person would pursue all possible avenues of remedy without minding this distinction too much.

Other patients who maintained a sharp distinction between the mental and physical disorders that underlay nervousness did so only provisionally, with the corollary that their conditions were clearly physical. Writing to Cabot about her son's recurrent attacks of indigestion, one woman offered a compelling argument about her previous physician's misunderstanding of the condition. Summarizing this other doctor's care in a careful typescript letter in 1914, she reported to Cabot that he was perplexed by her son's condition. He had suggested x-rays and then a surgical operation—in part, simply to diagnose the problem. This physician was, she believed, "inclined . . . to consider the illness due to [a] nervous condition." The boy's troubles, the doctor claimed, resembled symptoms he had met with "repeatedly in several children in his practice due to nervous condition." In countering this claim, the mother pointed out that her husband's family was "not disposed to nervous irritability, fears or melancholy." This other doctor's diagnosis of a nervous condition, to her way of thinking, shaded off into claims about simple anxiety, sadness, and fear. Her theory about her boy's problem was substantially different. She explained that her understanding was founded in knowledge of the boy's inherited predispositions from his father's family, whom I shall call the Batemans. The boy's father, a Bateman, also suffered from "attacks of indigestion like" her son's. So common was this condition in the Bateman family that they had taken to calling it "[Bateman] dyspepsia," she claimed.[45] She sought Cabot's opinion on the matter too, but suggested that the diagnosis was a familial dyspepsia that afflicted the Bateman stomach and not solely or merely the Bateman nerves.

These patients' concern that the physicality of their conditions be taken seriously found vindication in physicians' emerging views on nervous disorders like neurasthenia. As nervousness moved back out of the body, it entered more contentious territory. Physicians in Boston began using the diagnosis of neurasthenia almost as shorthand for the troubles that they encountered in excluding the bodily sources of physical symptoms. Searching for an explanation for a patient's troubles, a physician at Massachusetts General Hospital in 1906 complained in a note in the medical chart that his patient had "a typical neurasthenic complaint of pain in different parts of the abdomen, not constant."[46] The pain shifted and changed, eluding the determination of its exact bodily source. The diagnosis that this physician finally made was neurasthenia. The patient was "neurasthenic" and so was her complaint, making both perhaps equally perplexing. The diagnosis of neurasthenia in the early twentieth century became in the hands of some doctors an expression of frustration. Another doctor at MGH wrote in the chart in 1906 complaining about a similar case that there was "no diagnosis possible yet." His suspicion was that his patient "is certainly neurasthenic," making the process of diagnosis difficult. Neurasthenia became a term to designate patients whose discomforts and complaints belied their identifiable physical condition. As nervousness eluded diagnostic scrutiny, so the neurasthenic patient increasingly became someone who eluded medical advice, investigation, and control.[47]

It was in the nature of neurasthenia to frustrate diagnostic investigation. It had by 1900 become a condition that invited exposure as a cover for something else. Diagnosticians wanted to lay bare the real cases of tuberculosis, dysthyroidea, or dental disease that underlay it, or to reveal it as a mild version of a more substantive mental disease. Neurasthenia became almost by definition whatever remained behind after such attempts at diagnostic categorization. So one patient at MGH eventually diagnosed with neurasthenia in 1895 was initially noted to be "worried about heart." The physician who performed this patient's physical examination reported that it was "absolutely negative as far as the heart was concerned," an unusually emphatic conclusion for these records. Fifteen years later, a different physician at the same hospital recorded that another patient with neurasthenia also had "pain over heart," following up with a similarly definitive result that "exam of heart invariably negative."[48] In both cases, the physicians offered a decisive contrast between the inconsistency of the patient's complaints and conclusiveness of the physical examination. A patient with unfounded symptoms that shifted and eluded diagnosis could qualify as neurasthenic.

The anxiety of these physicians, suggested by their unusually definitive pronouncements, was rooted perhaps in a particular peril of diagnostic responsibility—a peril that grew in consonance with the growing weight of therapeutic obligation. If the heart pain really was from the heart, should not something be done? Recognition of this danger is evident in correspondence from another colleague of Cabot's. Writing to him in 1906, Dr. Hall introduced a new patient to Cabot. This patient's condition was a difficult one: "he has long had a gastric ulcer with neurasthenic symptoms supervening," Dr. Hall explained. The implications of the neurasthenic symptoms became clear as the account unfolded in the letter. The gastric ulcer, which had come to light during an operation this patient had recently undergone, had previously perforated and scarred over.[49] The event of the perforation of the ulcer had apparently gone unnoticed, however, obscured perhaps by neurasthenic symptoms. A missed diagnosis like this one was a major concern in an era when surgeons were claiming to be able to head off a disaster like a ruptured appendix, if caught in time. A young woman who was diagnosed at MGH in 1907 with inflammation of the bladder also received the diagnosis of neurasthenia. She had a variety of discomforts that seemed to perplex her physicians. The examining physician complained in her chart that "it is impossible to differentiate real pain from the neurasthenic background."[50] Neurasthenic pains and neurasthenic backgrounds simulated or obscured the physical sources of pain and could not serve as reliable guideposts to the identification of bodily abnormalities like ulcers or bladder infections.

On the wards at MGH in the first decades of the century, neurasthenia also became the diagnosis of a patient's resistance to medical authority and oversight. Patients identified as neurasthenic often seemed to their physicians to be as unreliable as their symptoms. So the diagnosis took on a distinctly pejorative tone, conveying a distrust and disapproval of its sufferers. "In spite of all efforts medical + psychic patient will not confess to feeling better," wrote one physician at MGH in the chart of a woman listed with neurasthenia in 1909. This doctor did not explain why he expected such a confession. Other physicians were more explicit about the source of their grievances. "Complains of not sleeping but may be seen sleeping," was the charge lodged in the chart against another patient diagnosed with neurasthenia at the hospital. Even more to the point, the same physician wrote on this patient's discharge: "Nothing suits her. No progress. Does not submit to treatment."[51] Resistance to medical definition and control seemed intrinsic to the condition of neurasthenia among these hospitalized patients. Suspicion about the vagaries of neurasthenia seems to have extended also to the vagaries perceived in the patients labeled with the problem. The private patient of Cabot's who expressed a worry that the diagnosis of neurasthenia would provide less substantiation for her condition than tuberculosis had a real basis for her concern.

The elusiveness of nervous disease and its resistance to medical definition and control seems to have inspired in some of Cabot's colleagues an eagerness to prove the absence of physical disease through their management of nervousness. Physical abnormalities might be sought exhaustively in people otherwise suspected of being nervous, as illustrated in the case, cited in chapter 3, in which a man who endured a long series of increasingly daunting diagnostic evaluations finally received from his physicians only a sedative tonic for his nerves. The emphatic pronouncements about the hearts of the neurasthenic patients at MGH seem to reflect a similar ethos. Cabot, although he denounced this diagnostic zeal in his colleagues, also on occasion demonstrated a similar enthusiasm to assert the absence of physical abnormalities in sufferers from nervous diseases. His desire to define and pin down nervous disorders encouraged the use of diagnostic tests for persuasive purposes, for example. Diagnostic technique could produce conclusive evidence about disease. But what use was a sophisticated diagnostic technique when it produced no such conclusive evidence? In Cabot's hands, diagnostic testing sometimes served primarily to convince a nervous patient about the psychic nature of his or her nervous disorder by excluding any physical disorder. This use seemed particularly important to Cabot, who argued in one essay, for example, that "there is no piece of medical science more clear cut and satisfactory than the power to reassure a person about an illness that he thinks he has."[52] Diagnostic technique provided a tool for creating such persuasive reassurance.

The most easily identifiable use of testing for persuasive purposes in this practice occurred among patients who suffered from what Cabot idiosyncratically described as the fear of disease. His office records are explicit about the management of what must have seemed to Cabot a relatively familiar problem. In most years between 1900 and 1915, roughly 1 to 2 percent of all patients diagnosed had problems attributed to their fear of disease. In 1910, one such man, who was in his twenties, visited the office expressing the concern that he had "smoked the pipe of a syphilitic man." Cabot's final diagnosis for him was "syphilophobia," and for the management of his condition, he received a Wassermann serological test for syphilis. When the test returned a negative result, it closed the case; diagnosis and treatment were complete, at least for the purposes of the medical record. Fear of disease was a condition both verified by the test and treated by its reassuringly negative results, while the documentation left an ambiguity about the order in which diagnosis and treatment occurred. The diagnosis of the fear of disease constituted a small but routine element of this practice. Cabot's diagnostic terminology for characterizing the problem typically cited the specific disease, from more common concerns like syphilophobia (as Cabot and others referred to it), cardiac apprehension, and fear of phthisis (tuberculosis) to more unusual conditions like fear of leukemia, fear of epilepsy, and fear of tabes (late syphilis). Each condition was amenable to identification and treatment in part through the appropriate negative diagnostic evaluation.[53]

In some cases, this persuasive reassurance seemed merely a by-product of the usual diagnostic process. In 1902, a young man came to the office with what Cabot would eventually diagnose as fear of tuberculosis. He was coughing occasionally, and another physician had already warned him about the possibility of incipient lung disease. The medical chart recorded a basis for this concern, describing him as "tall and long necked," commonly understood signs of a hereditary predisposition to the disease. The young man cooperated with an exacting search for tuberculosis through testing, including a normal temperature chart that he kept for the next ten days, a negative skin reaction to an injection of 10 milligrams of tuberculin, and two successive negative examinations of his sputum for tubercle bacillus—recorded as "Rx: sputum" in Cabot's chart. This cumulative negative evidence seems to have dispelled the concerns of both doctor and patient about tuberculosis. The young man wrote back a few years later saying that this visit had put his mind to rest on the question and reporting that he had "not since suffered in my lungs in any way."[54] The diagnostic process for tuberculosis offered what Cabot portrayed as a valid therapeutic effect, treating the fear of tuberculosis as a secondary consequence of the evaluation for disease, which is confirmed by his patients' letters.

Testing might create reassurance incidentally in this way, but there remained another, less easily defensible, but similarly persuasive, use for diagnostic testing. The self-conscious notation in these charts that diagnostic tests were an "Rx," a treatment, for the fear of disease reflects a rhetorical quality in the testing, conveying the impression of treatment, as well as the fact of diagnosis. In some cases, the distinction between Cabot's standard diagnostic intent and his simple persuasive intent seems blurred. How much evidence did Cabot need, for example, about the absence of syphilis in an apparently healthy man who reported only the possibility of a contaminated pipe?[55] The Wassermann test in that instance may have been more for reassurance of the patient than the physician. It was impossible in most instances to separate Cabot's intention to persuade from his intention to diagnose. However, these records did capture one instance in 1900 of an overtly manipulative use of diagnostic testing. Another patient who was similarly worried about tuberculosis, and "scared about long cough," was a "picture of health," as noted at the end of the first entry in his chart. His diagnosis was also "fear of phthisis," that is, tuberculosis. In this case the treatment was "Rx: sputa (not sent)."[56] The "Rx," the prescribed treatment, was to obtain a sample of sputum to send for examination for the infectious agent, the tubercle bacillus—a time-consuming process that Cabot delegated to an outside laboratory. In this case, however, the sample of sputum was not sent to the laboratory; no examination appears to have been made; and no result was recorded. But the diagnosis of fear of tuberculosis stood. The value of the test seems to have lain merely in creating the impression that it proved the absence of the disease. Cabot treated the fear of disease here through diagnostic testing as a form of placebo. By 1903, Cabot had denounced the intentional use of therapeutic placebos. No other overt instance of placebo diagnostic testing appears in these charts, consistent with his rejection of practices that he saw as nakedly deceptive. Perhaps in the early years of his practice, Cabot tried out and later discarded subterfuges that he had picked up from colleagues during training, including the use of placebos and of placebo diagnosis for deliberate persuasion. Cabot remained committed to the value of persuading nervous patients about the absence of disease but with time became more circumspect about his methods.

Psychoneurosis: Nervous Disease and Jewish Identity

Physical disease ideally manifested itself to physicians and patients alike in stable and predictable ways. The basic tenets of diagnostic reasoning and physiological treatment assumed these qualities. A patient indicated the site of a pain, and by following this clue, the physician could find the aberration and treat the underlying disease. Pathological alterations in the structure and function of the body, and their remedy, thus served both as a validation of the patient's pain and a verification of the diagnosis and treatment. Patient and physician could both perceive disease reliably from their separate perspectives. This model of physical disease emphasized the disease's naturalness, stability, and relative independence from the personal identity and social context of the sufferer.

It was a convincing model for medical practice, yet its strength derived in part from the ability to distinguish physical diseases from the many exceptions to the model, as illustrated by Cabot's struggles with general nervousness. Increasingly isolated from physical bodily sources, the nervous diseases were also contingent and dependent on the identity of the individual patient—to the extent even that different types of people manifested distinctly different types of the same nervous disease. The personal context of nervous disease and its subjective manifestations mixed inseparably with its identification, definition, and management in medical settings, making it difficult for a physician to talk about nervousness without talking about the life of the nervous patient. Cabot's earliest version of general nervousness as debility allowed for an eclectic handling of emotional, psychic, and physical disturbances. But as physical disorders became more exclusively the realm of objective evidence, nervousness was, in a parallel process, becoming segregated off with the personal, the social, and the contingent. The implications of this shift are especially evident in the clinical handling of nervous conditions in Jewish patients by Cabot and his colleagues.

Cabot and many of his Boston colleagues referred to the Jews as a distinct racial group. Their references to a separate Jewish race have a strange and even jarring sound. This oddness reflects in part ideas about Jewishness particular to their time, which we shall return to later. But in part it also reflects ideas about race that were in transition. The medical discussion of race in the early twentieth century carried an unfamiliar mix of implications. The familiar, and disturbing, questions about the existence of fixed, heritable racial differences were being raised forcefully. And the possibility of such differences drew widespread attention with the rising interest among American physicians in eugenic reform. But the use of race in this controversial sense shared space with more intricate and less stark medical conceptions of race as well. Nineteenth-century medical writing used the idea of race in a range of ways. By the end of the century, those who rejected the idea of immutable racial differences, like Franz Boas, and those who promoted egregious racial hierarchies, like Madison Grant, had staked out unambiguously opposing positions. They made it clear what people disagreed about when they disagreed about race. But physicians continued to fall back on an expanded conception of race that meant something closer to "a people," blurring putative hereditary distinctions into distinctions of national history and ethnic background. So the medical conceptions of a Jewish race in Cabot's heyday held in tension competing views about Jewishness alongside competing views on the nature of race.[57]

Ethnic identity infiltrated so deeply into the definitions of nervousness that Cabot and many of his Boston peers made the extraordinary identification of a separate Jewish racial variant of nervousness. Jewishness affected the expression of neurasthenia, debility, and psychoneurosis so fully in Cabot's practice as to create special subcategories of the disease peculiar to Jewish patients, which he designated variously over the years with terms like "racial neurasthenia," "Hebraic debility," "jew-neurasthenia," and "Jewish psychoneurosis." Physicians at Massachusetts General Hospital at the time also identified nervous conditions in their Jewish patients that seemed specifically racial in their terminology, giving rise to their use of a similar set of diagnostic terms. This concept of Jewish nervousness was an American inheritance from well-established European medical writings on the topic. But it also picked up and reflected home-grown medical reasoning about race, nervousness, and disease that extended widely through the American medical literature of the day.[58]

Jewish ethnicity and nervousness presented instructively analogous dilemmas for medical clinicians. Both Jewishness and nervousness had been claimed as categories of physical, bodily difference, yet these differences were also inextricably connected to personal history, customs, and circumstances. The sources of difference in both race and nervousness were ambiguous and open to a range of interpretations. So Jewish nervousness might seem to arise from within as an inherent difference; or it might exist as only the transient result of imposed circumstances in individuals who shared a common cultural history and experience. Even among Cabot's Boston colleagues, differences were evident in their inclination to make racial nervousness a stable, fixed characteristic of Jewishness. Their ability to disagree about the source of the condition may ironically have made it easier to agree about its existence and about its relevance to clinical action. Accepting the serviceable reality of Jewish nervousness, physicians debated its interpretation and implications.

Hyman Morrison writing in the pages of the *Boston Medical and Surgical Journal* as a senior medical student at Harvard Medical School in 1907 brought this debate squarely before his colleagues. He noted with evident distaste that doctors at Massachusetts General Hospital diagnosed separate Jewish nervous disorders, such as "Hebraic debility." To challenge this practice, he reported on his follow-up visits to the homes of fifty-one patients previously diagnosed at the hospital with a Jewish nervous condition. The topic was one that would make deep claims on the attention of soon-to-be Dr. Morrison. He would become one of the first Jewish physicians to break into the ranks of the Protestant attending physicians at MGH, and he became a foundational leader in Boston's Jewish medical community in the decades ahead. His critique of the concept of "Jewish nervousness" was an opening foray for him into a matter that occupied his attention in years to come as he faced deep-seated prejudices in the medical community where he built his career.[59]

In his article, Morrison accepted the existence of a higher rate of nervousness among Boston's eastern European Jewish immigrants. His major claim was simply that these nervous changes were transient, externally induced, and likely to improve. He rejected the position that he reported to be held by the Christian physicians at MGH that their Jewish patients differed fundamentally in their nervousness, that they had "symptoms peculiar to Jews, that they were more than mere neurasthenic symptoms." Nervous conditions were at root similar in Jews and non-Jews, Morrison argued. He insisted, against the teachings of his senior colleagues, that there was no such thing as a nervous disease "peculiar to the Jewish people."[60]

In making his case, Morrison joined, perhaps knowingly, a protest already well embarked upon by other medical scholars against anti-Semitic concepts in medicine. Maurice Fishberg in New York City, for example, argued against the existence of a separate, inferior Jewish "race," using medical data and observations to support his case. Fishberg was a physician and an anthropologist, a colleague and friend of the progressive anthropologist Franz Boas, who wrote extensively on this subject for both American and European audiences. Fishberg's 1911 book *The Jews: A Study of Race and Environment* followed on a series of articles in the American medical literature challenging the idea that differences in health and illness defined a fixed difference in the Jewish people. He was responding in part to a vigorous anti-Semitic medical literature of the times in asserting that "differences between Jews and Christians are not everywhere racial, due to anatomical and physiological peculiarities, but are solely the result of the social and political environment." Environment could change, making possible the smooth assimilation of new Jewish immigrants, which Fishberg hoped to promote. In Boston, Hyman Morrison, although he did not explicitly reject the idea of a Jewish race as Fishberg did, took a similar tack. He too argued that transient conditions of life had resulted in the medically observed differences in the expression of Jewish nervousness.[61]

But how deeply did the "social and political environment," as Fishberg characterized it, penetrate into specific differences in disease? And if it was not "anatomical and physiological peculiarities" that accounted for observed differences, then what might make a disease seem different among the Jews? Morrison addressed the question about the sources of Jewish nervousness head-on, although without resolving the more fundamental questions about the nature of racial difference. He speculated in his article in the *Boston Medical and Surgical Journal* that the Jews were "a highly imaginative people" whose terrible history of oppression, especially among the recent eastern European immigrants, had made fear "one of their keenest emotions." These emotional changes seemed to Morrison to be temporary, maintained as they were in part through cultural practices, so that "imitation and tradition play some part in the etiologies of these debilities." Yet he also allowed that habit and culture impressed changes in the physical expression of disease as well. Physical symptoms like constipation, commonly associated with Jewish nervousness by Morrison's medical colleagues, arose through the influence of a "general muscular hypotonicity which has become characteristic of the Jewish people during the ages of persecution." It was the nervous system principally that mediated these effects; so that Morrison cited another authority for the observation that the nervous system had come to dominate over the muscular system of the Jewish individual as a result of "the conditions of his existence." The nerves, in Morrison's assessment, mediated between bodily difference and the contingent social circumstances of a people. External conditions like persecution and oppression worked on the nervous system, creating the false impression of a stable, inherent difference in a separate race, and influencing the expression of physical symptoms.[62]

Morrison's brief response to his colleagues at Massachusetts General Hospital, although rejecting their conclusions, used Jewishness in a conventional kind of clinical argument. Race was a well-established and widely used construct in medical reasoning, from which the early twentieth-century physician might deduce a range of useful clinical conclusions. The observation that a patient was Jewish, for example, provided the physician with information about the chance of certain diseases arising. Nervousness was simply a special case of racial predisposition. Whether certain racial predispositions represented inherent and fixed differences or differences that were transient and imposed remained open to debate. Racial differences in the incidence and expression of diseases prompted speculation by physicians who held a range of differing views on the nature of human race—some with an explicit political and ideological basis.[63]

Cabot's own influential writings on differential diagnosis illustrate a fairly typical use of race in medical analysis, including observations on how Jewishness, which Cabot thought of as a racial difference, might figure in diagnostic reasoning. Throughout his major textbook on diagnosis, widely circulated in the second edition of 1912, Cabot set forth a panoply of examples on how to apply race to diagnosis. Claims to the effect that "tuberculosis is so common in the negro race that it is natural to suspect it whenever a negro is seriously sick," or that "tuberculous peritonitis should always be entertained, especially when the patient is an Italian recently settled in America" derived presumably from long experience with such associations.[64] These observations acted as rules of thumb for drawing probabilistic conclusions about diagnosis based on racial disparities in the occurrence of disease. Cabot continued in the same text that "clinical experience teaches that whenever a negress is sick and the symptoms are below the waist, fibroid tumor of the uterus usually turns out to be the diagnosis." The predispositions inherent in a particular group might be quite specific, and local, as for example Cabot's observation that "gonorrhea . . . infections are very rare in the young, unmarried Russian Jewesses of Boston."[65] Such associations suggested behaviors and local customs as the link between racial difference and disease, but Cabot provided little overt explanation or discussion about the influence of race on disease.

Nervousness too was a condition that showed differential susceptibility by race, but race was only one among a larger set of factors that figured into diagnostic reasoning. A hospitalized woman, also discussed in the textbook, was reported to be a "Swedish housemaid of twenty-five" who complained of long-standing backaches. Cabot speculated that "functional, neurasthenic, or hysteric affections of the spine are naturally suggested by the long duration of the symptoms, by the age and sex, and by the absence of fever," although, in this instance, an x-ray revealed that her condition was in fact vertebral tuberculosis. Race numbered among a list of characteristics that the physician should consider in estimating the probability of disease. Racial or ethnic identity as Swedish or Jewish had pertinence in estimating the likelihood that a patient was suffering from a nervous disease like neurasthenia.

In offering these associations between race and disease, Cabot handled race in a manner similar to Morrison's and typical of this period, although they disagreed fundamentally about the nature of Jewish nervousness, as we shall see. Cabot, like Morrison, was hesitant to settle on any single mechanism linking racial identity or circumstance with disease. He did not explain the cited connections between race and tuberculosis or gonorrhea, or other similar associations, for example, even when such putative explanations were discussed elsewhere in the medical literature. Diagnostic reasoning using race did not require examination of the underlying cause for these associations. Cabot noted that hydatid disease, for example, appeared more commonly "in association with sheep and sheep-dogs, especially in Greece, Australia, and Iceland,"—without citing the microbiological mechanism invoked in etiologic discussions of the disease. Nor does Cabot's reader discover, in considering a patient's job, why "work in a rubber factory often produces a stubborn type of general debility."[66] Race and attributes like social status or occupation affected the predisposition to disease in manifold ways, and their use in clinical reasoning did not commit the physician to a position on the cause, whether inherent or imposed. Morrison shared with Cabot a similarly common means of handling racial characteristics in the clinic. Both physicians reasoned about racial differences in disease in a way that assumed an unstated and unexamined norm. White, Anglo-American, or Yankee racial predisposition seemed to represent a baseline from which people of different race diverged. Predisposition to disease without a racial character seemed to presume a reference to a patient who was white—although Cabot and Morrison were again both typical in avoiding analysis, or even mention, of this assumed standard.

The sources for this logic of racial predisposition were doubtless complex. But a hesitation to analyze disease further into racially distinct categories had one evident basis. Racially distinct categories of disease might suggest distinct variability in the natural patterns of disease. Yet the acceptance of specific, unified types of natural human disease marked a hard-fought victory for the medical science of the early twentieth century. Having left behind a nineteenth-century nosological system in which "typho-malarial fever" might split the difference between typhoid and malaria, or "cholera infantum" shade gradually over into cholera morbus, physicians in the early twentieth century demonstrated restraint in positing the existence of distinct racial variants in natural diseases. This hesitation made Jewish nervousness, for example, such a critical exception. If diabetes, for example, was a unitary natural phenomenon, then it should be identical in its essential manifestations in every human being. Individuals with different susceptibilities might respond differently to the disease. But the idea that diabetes itself differed systematically in its natural course in different races cut against the grain of the etiological reasoning that was being so fruitfully applied in early twentieth-century medicine.

In considering a disease like diabetes, physicians turned to nervousness itself as a valuable linkage accommodating the unitary nature of bodily disease with its different racial expressions. Nervous predisposition was a serviceable conduit for the influence of racially determined differences

on unitary physical disease. Diabetes drew special attention for its putatively higher rates among Jewish patients. The high rate of diabetes
seemed never to imply to physicians a racially distinct form of the disease—
there was no "Jewish diabetes."[67] Even the vague suggestion that some
physical difference contributed to a racial disparity in disease met with
strong rebuttal. It was not, as one editorialist suggested, an "admixture
with Indo-Germanic blood" that was responsible for these racial differences, but rather "nervousness, overstrain, overwork and cares," that were
"factors in the etiology of diabetes" among Jews.[68] Nervousness was only
one consideration in medical speculation on Jewish predispositions to disease; physicians postulated variously that ritual circumcision, ghetto
crowding, marital customs, or dietary laws predisposed the Jews to different rates of cancer, tuberculosis, diabetes, or syphilis. Nervous predisposition sometimes played a fascinating role, however, in mediating these
differential effects.[69]

Remarkably similar connections between race and nervous temperament occurred to physicians contemplating putative racial differences in
syphilis. There was reportedly a low incidence of early syphilis among Jewish patients, an editorialist in the *Journal of the American Medical Association* noted in 1900. Paradoxically, this low incidence partially masked a
disproportionately higher incidence of the late neurological complications
of the disease. Nervous disorders supplied a convenient link between racial
characteristics and this disparity in disease expression. These apparently
higher rates of advanced neurological syphilis, the editorialist conjectured,
were referable to greater Jewish propensity to nervousness. It was, the
author asserted, "a well-known and, we believe, generally admitted fact that
the Jewish race is neurotic to a greater degree than others." Higher levels of
neuroses combined in Jews with the "strain of mental activity" to create a
special vulnerability to the rarer advanced forms of syphilis among members of a group relatively free from the otherwise more common early
stages of the disease.[70]

Clear pitfalls threaten an effort to generalize from theorizing about a
Jewish race in American medical literature to the discussion of other racial
disparities in the expression of disease.[71] Yet a remarkably similar set of
speculations connecting race, disease, and nervousness appears in the
medical literature concerning African American racial characteristics. In
syphilis, it was thought that a reverse of the paradox seen in the Jewish race

applied. The relative rarity of advanced syphilis among African Americans contradicted the expectations of white physicians about a race that they typically vilified as "syphilized," with a high incidence of the disease in its common early stages. Dr. E. M. Hummel, writing in the pages of the *Journal of the American Medical Association* in 1911, reflected on this conundrum. He aired skepticism that specific physical differences between white and black races created different expressions in the disease, readily dismissing a hypothetical difference in the anatomy of the skull as inconsequential, for example. Instead, it was "temperamental and nervous defects" that accounted for the higher incidence in the white race. He shared this view of the problem with contemporaries like Dr. Thomas Murrell, who in writing on "Syphilis and the American Negro" a year earlier in the same journal asserted that syphilis in blacks "is not to any great extent different from the disease in the whites." Murrell too wished to account somehow for the seeming disparity in the incidence of advanced disease with the higher than expected incidence in whites relative to blacks. Nervous temperament provided a crucial link for both physicians that avoided a racial distinction in the physical nature of the disease. The "highly civilized" white race, as Hummel phrased it, suffered from the nervous strains of their complex social world, making them more vulnerable to the effects of advanced syphilis than their "primitive" black peers.[72] Nerves linked social differences between the races convincingly to differences in disease, without requiring a physical difference in the expression of a stable disease like syphilis.

Convinced of the universality of biological disease, even American physicians who were inclined to accept fixed physical, hierarchical differences between races still might reject the notion that simple physical diseases differed fundamentally by race. Nervous predisposition based in mental states or sensibilities lent themselves to array along analogies of hierarchy, for example, from infant to adult, from primitive to cosmopolitan, from quadruped to human, or from sensuality to intellectuality. These means of sorting and describing racial differences might then apply to the analysis of racial differences in the incidence of disease. Such views were entirely compatible with the belief in fundamental hierarchical differences among races, in spite of the continuities in natural human disease. Writing in 1908, Dr. S. T. Barnett communicated the observation that appendicitis in people of African descent had "pathological findings so characteristic" of the disease in the white race as to be indistinguishable. He presented this result as worthy of special note, imparting the sense that he suspected there were other fundamental differences between the races. An 1893 tract on pregnancy in the "Negroes of South Carolina" noted in passing that the average measurements of the pelvis differed significantly between the white and black races.[73] These differences affected childbearing. Yet in considering the specific diseases of pregnancy, these cited bodily differences were not applicable. For example, higher rates of eclampsia among African Americans might be explainable "by the imitation of so-called civilized habits of the Whites." Once again, a difference in external conditions and circumstances translated into a difference in disease through the intermediary mechanism of nervous and mental habits and predispositions.

Nervous disorders and predispositions enabled some medical theorists to preserve a sense of the biological fixity of diseases like syphilis and diabetes, while acknowledging their different incidence in different races. The nervous system provided a useful link between fixed disease and the fluid effects of racial circumstances or vaguely defined inherent differences in "racial character," "sexual energy," or "moral tendencies." Race and nervousness, in fact, occupied somewhat analogous roles in the application of objective, scientific analysis to medicine. As physicians developed their ability to handle a stable array of natural human diseases, they shaped methods of clinical reasoning that accommodated both qualities of race and nervousness similarly. Whether understood as inherent or imposed, both racial and nervous characteristics often took the role of mediators between the natural biology of human disease and the contingent effects of personal habits and social conditions.

Nervousness and race, in these intermediary roles, could combine in a condition like Jewish neurasthenia, emphasizing the ways in which nervousness differed from bodily disease. Nervousness required the physician to account for the details of personal and social life in defining and managing it. A middle-age woman, likely Anglo-American, judging from her name, whom Cabot saw in April 1913 appeared agitated and teary on her first visit to the office, reportedly after "an altercation c[um] husband just before she left." He found her blood pressure elevated at 155 and her pulse irregular. Although he identified her "hypertension" as a problem too, he chose to focus on what he figured as her primary diagnosis, psychoneurosis. In the cases of some nervous patients, as we have seen, Cabot parceled out their therapy to psychotherapists and lay healers. But he treated others, like this woman, with his own combination of discussion, supportive encouragement, and advice. Cabot left scanty notes on subsequent visits, but he seems to have determined that painting helped her to quiet her nerves. He wrote out a kind of prescription, which she later returned to him with annotations, advising about her daily routines and recommending that she "paint something every day whether New York is stupid or not."[74] Scraps like this one suggested that Cabot's treatments involved more conversation and personal inquiry than made its way into the medical chart. But patients in their correspondence indicated that what constituted talk therapy for Cabot ranged across a great variety of personal issues. The pursuit of talk therapy for Cabot often left him to record rather cryptic data in the medical charts of his nervous patients. The initial chart entry for one patient diagnosed with psychoneurosis concluded with the notes:

> Has cared for a woman s[ine, i.e., without] physical. Physical punishment attractive. Has look[ed] on sex relation as indecent & it is therefore that he didn't win his girl. No reproach because he never let in anything. Sensual = sexual for him. Animal comes s[ine] halo—by itself. What is purity?
> Rx = Go ahead
> $10

A reader of the chart is left to wonder what it was that Cabot had prescribed with his permission to go ahead. Clearly, it was not a medication. In nervous disease, the boundary where the personal and social left off and the disease took over remained both indeterminate and flexible.[75]

The personality and life of a nervous patient infiltrated and defined nervous disease, so that the social and the pathological were linked more tightly than elsewhere in the nosology of the clinic. The handling of neurasthenia by another of Cabot's New England colleagues, Dr. Herbert Hall, demonstrates how variants of nervous disease, like Jewish nervousness, exhibited the social characteristics of the patients that they afflicted. Nervousness in the upper classes of society, in the context of Hall's medical practice, appeared distinctly different from the nervous condition among laborers. Herbert J. Hall was among the early leaders in organized occupational therapy in America, and served toward the end of his life as one of the first presidents of the American Occupational Therapy Association. A graduate, like Cabot, of Harvard Medical School in 1895, Hall also apprenticed briefly at Massachusetts General Hospital, and he started in private medical practice in nearby Marblehead. Hall's main interest was the treatment of neurasthenia through "work therapy." To this end, in 1904, he created the Devereaux Workshops, a sanitarium in Marblehead where patients suffering from nervous fatigue could recuperate through a program of graded manual craft work. Hall's success with the programs at Devereaux inspired him to put out a series of publications on the nature and care of nervous debility through what would come to be known as occupational therapy.[76]

To Hall, neurasthenia was a patently different problem in people of different economic backgrounds. In his first comprehensive book on the treatment of nervous disorders through work therapy, Hall saw fit to include separate chapters on "The Well-to-Do Patient at Work" and on "Treatment for People of Small Means."[77] Economic circumstances, as they meshed tacitly with gender, mattered directly in the care of neurasthenia, but not simply in creating practical limitations for people who could not afford an extended leave in Hall's seaside sanitarium at Marblehead. In different classes of patients, the disease of neurasthenia itself seemed altered, beyond the practical requirement to treat it within the patient's means. The wealthy patients, especially the women, suffered from a neurasthenia that was characterized by depletion of nervous energy, brought on in "the rush of social engagements, the stimulation of travel, the search for amusement." Work treatment offered an answer to their particular needs. Occupying hands and mind in gentle, productive activities restored nervous tone and balance, since "when the nervous invalid gets down to honest work with her hands she makes discoveries. . . . She learns something of the dignity and satisfaction of work and gets an altogether simpler and more wholesome notion of living."[78] This appropriately directed work-therapy acted as an antidote to the stress of a hurried and aimlessly nonproductive existence among female nervous invalids of a certain class. Work meant a shepherding and restoration of vital nervous energies.

"Honest work with her hands" was presumably not a novel experience, however, for the neurasthenic woman who did not come from the world of the leisured gentry. Hall intriguingly portrayed neurasthenia and its treatment as entirely different among patients who habitually engaged in manual work. His distinctions had obvious strategic implications. He needed to differentiate clearly between his carefully orchestrated therapeutic program of craft work and the manual labor that his patients pursued in their usual employment—otherwise what was the point of therapy. He emphasized the unique therapeutic qualities of supervised handicrafts in healing neurasthenia among "People of Small Means." Whereas craft work conferred distinct spiritual and characterological benefits on the wealthy, for those less well situated the same craft work offered physical and bodily therapeutic advantages. Handiwork served to restore bodily tone in "muscular fatigue" and to strengthen the support for the lax internal organs—"the condition called enteroptosis," which Hall believed was often implicated in nervous exhaustion among this group of working people. The nervous problems of neurasthenia among the laboring patients had a more physical character. Therapy, of course, yielded emotional benefits that complemented the physical. Craft therapy for the poor worker, Hall argued, "brings back confidence and strength and renews wasted nerve and muscle tissue." Wasted muscle tissue and lax internal organs did not figure into Hall's reasoning about neurasthenia among the wealthy, where the condition seemed altogether more psychic in its manifestations. Among patients of modest means, however, the disease was defined in important ways by its physical properties. Dr. Hall's prescription for work thus took on a different significance in treatment according to social class.[79]

Elements of a patient's race, gender, and social status, in the loosely defined terminology of these medical writings, seemed at times to affect the expression of nervousness so much that the disease itself was changed. The stable, natural existence of a disease, however, provided a reliable basis for the physician's influence and control over care. The mixture of the changeable and contingent conditions of human social life with nervousness also marked a dilution of this natural medical authority, which Cabot particularly sought. The kind of influence that physicians could assert over the nervous patients differed. The handling of Jewish nervousness by Cabot and his peers suggests that racial hierarchies served in part to reinforce medical control over a difficult form of illness. In an area of medical practice where knowledge about the natural properties of disease yielded little advantage, knowledge about human racial differences seemed to act as a substitute basis for clinical authority.

Jewishness in Cabot's practice impugned the reliability of the patient and had special relevance to the management of nervous patients. One of the challenges of nervous patients was their propensity to mislead the physician about the physical source of a problem. A physician's ability to ascertain the reliability of a patient was thus a valuable tool for defining and controlling nervousness. So the deceptiveness ascribed to Jewish patients by Protestant physicians like Cabot helped in part to anticipate and partially account for the elusiveness of their disorders. Cabot worked out these premises in his practical advice to his peers. In his popular textbook on differential diagnosis, he cautioned medical readers about patients who spoke of having "burning pains." The term was, he advised, "used by the Jews far more often in describing their symptoms than by any other race, and, as a rule, patients who use this term turn out to be free from organic disease." Such advice originated with his experience in the clinic. In his private office records, Cabot noted, for example, that a patient was afflicted with what he called a "jew pain," linking this unusual symptom to the diagnosis of a nervous condition. Such characteristic symptoms and pains were diagnostic clues to nervousness, seeming reliably to predict how unreliable the patient would be.[80]

Cabot's peers in Boston were disposed to use similar assumptions about Jewishness in managing nervous conditions. A patient at Massachusetts General Hospital whose physician complained that "it is impossible to differentiate the real pain from the neurasthenic background," was also identified in the same entry as a "neurasthenic Jewess," a recent Russian immigrant. Although thinking about Jewish race by the first decades of the twentieth century was beginning to realign in ways that would eventually incorporate Jews as a religious or ethnic group within a larger white race, this shift took place against a background of persistent anti-Semitism in public policy that became especially vehement in debates over immigration law early in the century.[81] Boston's medical world in the day offered Jewish patients a harsh reception. In his textbook, Cabot went on to show how his colleagues at the hospital sometimes countered the anticipated unreliability of their Jewish patients with a manipulative approach to diagnosis. He described the case of "a Jewess of thirty" with terrible belly pains under the care of his colleagues, most likely at MGH. In diagnosing a nervous condition underlying her pains, these physicians were, Cabot observed, "influenced by the fact that this patient's pain was greatly improved by the 'lie cure' (injection of sterile water, mistaken by the patient for morphine.)" A deceptive placebo treatment seemed to unmask nervousness as the source of pain rather than some bodily affliction.[82] This approach to diagnosis was one that Cabot rejected as liable to damage the physician's legitimacy. Perhaps he would have been willing to rely instead on his assumptions about the racial characteristics of the patient as sufficient evidence of nervousness.

Richard Cabot practiced and lived in a setting rife with well-documented, if infrequently discussed, anti-Semitism. Boston lagged behind New York as a prime destination for Russian and other eastern European Jewish immigrants at the turn of the century, but the dominant Jewish identity of Boston's West End was a distinct element of the political landscape. The West End lay close by Massachusetts General Hospital, making its Jewish residents natural clients for the hospital's outpatient clinics and wards. These circumstances were greeted at times with open resentment by a largely Anglo-American, Protestant medical staff, who complained that even those Jewish patients who should be ineligible for charitable care "know the lies to be told and the appearances to be put on to gain admission" to charitable service, as one trustee complained. Cabot's own family bloodlines laid down an additional crucial background, for nearly a century providing MGH with many of its medical and civic leaders. The Cabots were among the so-called Brahmins, so named by another famous Boston physician, Oliver Wendell Holmes, numbering among Boston's older, elite cohort of Anglo-American families with deep claims on the city's civic and political life. The Cabots, Lowells, and Lodges, among a short list of others, constituted an older mercantile class of established wealth. By Richard Cabot's day, changes in Boston's political constituencies made this older caste seem vulnerable or perhaps even humorous in the twilight of its influence. In 1910, John Collins Bossidy toasted

> . . . *good old Boston,*
> *The home of the bean and the cod,*
> *Where the Lowells talk to the Cabots*
> *And the Cabots talk only to God.*

By the early 1900s, the question might reasonably be asked as to whether anyone else was straining to listen in. Although Brahmin influence had by the early twentieth century lost some of its heft in Boston's political life, and in the city's collective imagination, figures like Senator Henry Cabot

Lodge remained from the lineages to make clear that such influence was no laughing matter. Lodge's well-publicized anti-immigrant views demonstrated a distinct revulsion for America's recent Jewish citizens.[83]

In this context, Boston's Protestant physicians demonstrated a frustration with Jewish nervousness that served to highlight their contrasting satisfaction in handling physical disorders. The difficulties of evaluating nervousness might be redeemed by diagnostic perseverance, so that by acknowledging racial propensities to unreliability, the physician could transcend these obstacles to find the source of "real pain." In a case of diabetes, for example, a physician might reassert the clarity of medical control even in a setting where nervousness seemed to predominate. Racial hierar-

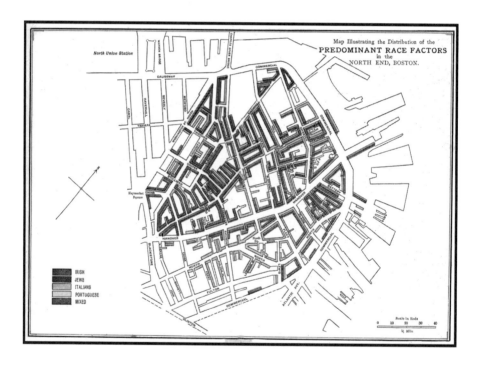

A traditionally Jewish neighborhood around Salem Street, Boston, ca. 1900. Salem Street addresses appear in Cabot's patient records. Massachusetts General Hospital is about half a mile to the east of Canal Street. From R. A. Woods, ed., *Americans in Process* (Boston: Houghton, Mifflin, 1903). Courtesy of the Trustees of Boston Public Library.

chies in reliability assisted in this process. Cabot's textbook on diagnosis outlined the instructive case of a "Jewish married woman thirty-four years old," who was seen in a consultation in 1907 for problems with headaches, vomiting, and weakness. The patient had, by Cabot's account, the kinds of problems found in "innumerable Jewesses of this age without our being able to discover any more definite cause than their self-starvation and a psychoneurotic constitution." In this instance, however, the stalwart pursuit of a physical diagnosis revealed a different underlying problem. "Five and a half per cent. of glucose was found in the urine," he reported definitively. This patient had diabetes mellitus, and this, rather than nervousness, was the source of her manifold symptoms. With careful attention to diet, her urinary sugars, and symptoms, soon cleared up.[84] By noting that there were "innumerable Jewesses" who suffered from superficially similar problems, Cabot emphasized the diagnostic challenges in Jewish nervousness, as a mask for other substantive medical problems.

Knowledge of race complemented the physician's knowledge of natural disease in establishing the boundaries between nervousness and bodily abnormality. By recognizing a Jewish tendency to exaggeration, the diagnostician might overcome it in identifying a physical disorder. Such was the case with "a highly neurotic Jewish boy of eighteen" whom Cabot also described in his textbook. The principal difficulty, to Cabot's way of thinking, was that "any one [so neurotic] especially if he be one of the Jewish race runs a considerable risk of being falsely accused or falsely suspected of being 'merely a neurotic'." Instead, a careful diagnostic evaluation established that this young man had a kidney stone, and that this was the cause of at least some of his troubles. He was neurotic, but not only neurotic. Knowledge about racial hierarchies in reliability helped the physician in asserting this distinction.

A patient's racial identity situated him or her within a hierarchy of reliability in reporting symptoms. If Jews were highly unreliable, other groups seemed slightly less so. Also in the text on differential diagnosis, Cabot outlined the case of "[a]n Italian laborer of twenty-four" with right flank pains. In dismissing a nervous disorder as a possible source for this man's trouble, Cabot noted that "[y]oung Italian laborers rarely suffer from functional neuroses," in contrast perhaps to the young Jewesses considered above. He then continued, qualifying this generalization: "I have once known a case somewhat similar to this in which the patient turned out to be a malingerer, but he had obvious reasons for his lies, while this patient has none."[85] This observation suggests an underlying moral evaluation of race. When Italian laborers reported pains for which no physical source could be found, they at least tended to have clear-cut motives, Cabot suggested— making the assessment of their reliability that much easier. These Italian laborers manifested an earnest dishonesty, in which they were themselves quite clear about the absence of real symptoms. They had a reason for their lies. Thus, in this case, Cabot immediately rejected a functional neurosis— an unintended, neurotic deception—and segued into consideration of malingering based on an earnest intent to deceive.

Considerations of race, gender, and occupation all carried weight in working out the differential reliability of nervous patients, especially the reliability of their reports on subjective symptoms. In his textbook on diagnosis, Cabot reviewed the case of a "[J]ewish housemaid of twenty" who suffered from backaches. She, like the Swedish housemaid mentioned earlier, eventually received the diagnosis of tuberculosis of the spine, but a key consideration in her case was that "a functional neurosis is not likely in a girl who keeps steadily at work, although in constant pain."[86] This patient's identity as Jewish seemed to trigger at least a consideration of a nervous disorder as an explanation, although racial identity did not overwhelm the significance of her work habits. Reasoning about nervous disorders required this kind of subtle calculus of reliability, based on knowledge of common social hierarchies that located Swedish and Jewish housemaids with some precision.

Racial and social hierarchies offered some leverage in evaluating the nervous patient. Yet the power conferred by this knowledge of social conditions appeared distinct from, and less reliable than, a technical knowledge of disease. When Cabot examined his assumptions based on race explicitly, it was mainly to question their validity. He connected the Jewishness of patients to a distortion, not only in their reporting, but in his own perceptions as a medical observer. Addressing an audience at the New England Hospital for Women and Children in Boston in 1906, he described his struggle against the influence of his anti-Semitic prejudices on his medical practice. His assumptions about Jewish patients seemed to him at times a hazardous distortion in clinical perception. He worried that when he saw a Jewish man in his examination room, "I do not see *this* man at all. I merge him into the hazy background of the average Jew." The barrier was one of prejudice; Cabot found that instead of the individual man, he saw only "a Jew, a nervous, complaining, whimpering Jew, with his beard upon his chest and the inevitable, dirty black frock-coat flapping about his knees." The trouble was to discern the patient's true condition through the lens of anti-Semitic prejudice: "I never saw *them* [my Jewish patients]," he stated, "but only their ghostly outline, their generic type, the racial background out of which they emerged."[87] Cabot attempted to repudiate this prejudice as an obstacle to good medical practice, although he continued to register it in his diagnostic nomenclature and in his practical advice to his medical colleagues.

One characteristic of an authority based on knowledge of bodily disease that may have appealed to Cabot was its seeming transparency. Cabot could present the results of an x-ray or a blood count when it helped to support diagnosis or treatment. An authority based in social hierarchies might be less readily opened to examination. The influence of a Christian physician over his Jewish patients might reflect differences in social power. But could it reliably generate and sustain these differences? Cabot developed a complicated relationship to Boston's Jewish medical community during his career that seems illustrative of this conundrum. Although Cabot demonstrated significant antipathy toward his Jewish patients, both in print and in practice, he made himself into a powerful ally of Boston's Jewish physicians, leading Boston's medical elite in their support. In 1907, Cabot became president of the medical staff in the city's first Jewish hospital, the Mt. Sinai, serving in that capacity until 1914. Mt. Sinai was established in part to give Jewish physicians who were shut out of other hospitals a place to practice. But other avenues were opening up as well. Massachusetts General Hospital soon admitted its first Jewish physician to the medical staff, in the outpatient service, and Cabot was credited with securing this concession.[88]

Dr. Hyman Morrison experienced the ambiguities of Cabot's relationship to the Jewish medical community firsthand. He was a student of Cabot's at Harvard Medical School when he undertook his study of Jewish nervousness, published in the *Boston Medical and Surgical Journal*. In a signed copy of this paper that Morrison deposited in the archives of Harvard Medical School, he indicated that Cabot had initially suggested the project to him—probably soon after he gave his speech at New England Hospital denouncing his own prejudices about his Jewish patients. The anonymous hospital physician whose views Morrison criticized in his paper may well have been Cabot himself. Cabot implicated himself in the very practices that he inspired Morrison to investigate, and that subsequently came under Morrison's critique. Was Cabot looking for a confirmation from his Jewish colleague that his reasoning about the race was sound? It is difficult to imagine exactly how Morrison understood Cabot's role in initiating the study—whether as a mentor or a provocateur. Later, it was again Cabot who helped direct the campaign for approval when Morrison became only the second Jewish physician admitted to the staff at MGH.[89] Cabot worked hard to advance the position of his Jewish colleagues, even as he identified their race a marker of deceitfulness among his patients.

"Not Ill": Nervous Women and Anxious Men

Hierarchical assumptions about race functioned quietly in Cabot's clinical practice, acting to shore up the physician's ability to manage nervousness, but without yielding a self-consistent and sustainable logic in their use. When Cabot subjected racial hierarchy to scrutiny or trial, at least for Jews, it seemed to lose its legitimate power to inform practice. Cabot seemed unwilling to consider, for example, how Dr. Morrison's reliability differed from that of his patients of the same putative race. The authority of a male physician over female patients operated in an analogous fashion in Cabot's practice. Gender helped to sort patients according to their nervousness, serving as an explanation similar to race in its application to clinical reasoning. Yet Cabot seems to have been even more reluctant to consider the validity of distinctions according to gender, seeming to use gender silently as a mechanism for controlling nervousness, without any open discussion or consideration of its role or purpose.

In each of the years from 1909 through 1915, during the height of Cabot's reliance on the diagnosis of psychoneurosis, between 9 and 21 percent of all new patients received this diagnosis. Among Cabot's patients with psychoneurosis, fully 78 percent were women. During the same seven-year period from 1909 to 1915, Cabot also diagnosed from 2 to 7 percent of his patients as being "not ill," a diagnosis that he seemed to introduce coincident with the increased reliance on psychoneurosis. Among these patients who were "not ill," only 42 percent were women. Both these

TABLE 5.2   COMPARISON OF AGE, SEX, REFERRAL STATUS, AND FEE CHARGED, ALL
PATIENTS IN THE RANDOM SAMPLE AND PATIENTS IN THE SAMPLE WHO RECEIVED A
DIAGNOSIS OF DEBILITY, NEURASTHENIA, PSYCHONEUROSIS, OR NOT SICK

	All patients in the random sample	Patients diagnosed with a nervous disease	Statistical significance
Number	200	73	
Average age (years)	38.7	29.3	$P < .001^a$
Female	104	41	ns[b]
Means of contact			$P < .01^c$
Consultation	44 (22%)	9 (12%)	
Independent	118 (59%)	56 (77%)	
Doctor referral	38 (19%)	8 (11%)	
Average fee	$13.00	$9.40	ns[a]

Note: Missing data in population = >20%; ns = not statistically significant.
[a]Two-tailed t test.
[b]Z test.
[c]Chi-square test.

figures differed significantly from the overall fraction of 52 percent women
in the random sample of patients (table 5.2). People diagnosed with psychoneurosis were much more likely than the average patient to be female;
and people who were diagnosed as not ill were much more likely to be
male. The overwhelming majority of patients diagnosed as either not ill or
psychoneurotic had one thing in common: they suffered from bodily distress and concern but had no readily identified physical abnormality. They
were nervous. Gender was a feature that seemed to sort nervous patients in
Cabot's practice into those who should be confirmed in their nervousness
and those who should be persuaded of their physical health.

A similar pattern of distinction by gender applied to the earlier period
of Cabot's practice, between 1900 and 1904, when the diagnosis of debility
predominated. Fully 70 percent of Cabot's patients with debility were
women. The counterpart to the diagnosis of "not ill" from this earlier

TABLE 5.3   WOMEN AS A PERCENTAGE OF PATIENTS
DIAGNOSED WITH NERVOUS DISORDERS OR AS NOT SICK

Debility	70% (112/159)
Fear of disease	25% (13/51)
Neurasthenia	60% (130/218)
Psychoneurosis	78% (333/429)
Not sick	42% (67/157)

period was "fear of disease," as can be seen in the figures in table 5.1. From 1 to 3 percent of all patients in the years 1900 to 1904 were diagnosed as suffering from the fear of a disease, while almost no patients during this time were diagnosed as not ill. Among patients diagnosed with the fear of a disease, only 25 percent were women (table 5.3). This division by gender remained similar in both the early and late phases of practice. Between 1905 and 1909, Cabot relied more heavily on the diagnosis of neurasthenia, only rarely in this transition making a diagnosis of either "not ill" or "fear of disease." Neurasthenia was a diagnosis that applied to about 60 percent women, very similar to the 56 percent of women among the nervous patients in the random sample. (Compare figures in tables 5.2 and 5.3). Many individual patients proved an exception to this generalization. But overall, the pattern was quite distinct. Cabot tended to find that his nervous male patients needed to be convinced of their fundamental healthiness. The women suffered from actual nervous disease.

Cabot offered no published recommendations or indications about how to handle this characteristic relationship between nervousness and gender in practice. He did repeatedly echo the common wisdom of his day that women suffered more from nervous disorders than men. The corollary that men should more often be dissuaded of nervousness remained a seemingly unspoken tenet of practice. Whether he was even aware of the influence of gender on his diagnostic habits is unclear. The authority of a male physician over women patients was again a kind of authority that did not lend itself to examination as an element of proper medical management. Cabot appears in this case to have been simply unconcerned that this inclination held sway.

For Cabot, and for many of his medical colleagues, the social and personal context of physical disease mattered. Context did not redefine the natural course of bodily disease—as it seemed to with the nervous diseases—but it did alter the way the bodily disease affected the individual patient. What was most striking about the social and personal contexts of physical disease was how they could be stripped away from the disease, at least hypothetically. A case from Cabot's clinical records provides an example. An administrator at a major educational institution, a woman in her forties, came to 190 Marlborough Street in December 1910 with problems that were diagnosed as asthma and bronchitis. Her doctor at home had written to Cabot seeking support for his recommendation for her to leave her stressful occupation in the harsh northern winters for a healthier way of life. Her breathing problem was worsened by the local climate, he believed, and by the stressful nature of her work there. Attention to these factors would effectively treat it. Cabot did not disagree; but inspired by other considerations, he set out to instead overcome these concerns by treating the bronchitis directly. First, he assured himself as to the nature of his patient's condition by a series of microscopic examinations of her respiratory secretions. He then prescribed a rigorous, month-long course of pharmacological and physical therapies for her breathing, in preparation for her return to her important administrative position. Cabot would have been the last to argue that her life circumstances were inconsequential to this disease. But he planned to treat her condition in spite of the exacerbating circumstances—not ignoring them, but attempting to supersede them through direct intervention in bodily function. After this lengthy course of prescribed treatments, the administrator returned to her position in the northern climate.

The question for Cabot was in part about the nature of medical influence over her care. Her home physician had written supporting a change of environment and job, but he seemed compromised in his recommendation. He explained confidentially to Cabot that this woman's home institution, where he also worked, planned to seek her resignation for reasons unconnected to her health. He argued that the stress of being asked to leave would imperil her health further, and so he had shrewdly advised her "to send in her resignation on the ground[s] of her health in order that she may never know the action of the board."[90] His advice about her health in this case conveniently coincided with what he knew about the secret intentions of the supervising board. His medical recommendations would anticipate and covertly support their intention to fire her. It may have seemed like a reasonable opportunity to link medical influence to the powerful ally of the local board, who would inevitably win in any case. The home doctor was acknowledging the complex circumstances of her job and also taking advantage of them—with a potential benefit to her health, as he explained it.

Cabot was understandably resistant to this conspiracy of medical advice. In treating her lungs, he hoped to provide effective medical care in spite of, or even in opposition to, the circumstances of her life. She would face opposition at home with the best advantages that her well-treated lungs could provide. She went back to her job and wrote to him later supporting his decision, since "at the distance of three years I am able to see clearly that it was my privilege to know that I had critics . . . and that you were right to let me return to find it out." Whether Cabot told her about the conspiracy or just let her discover it for herself remains unclear. Unfortunately, Cabot had less ability to leverage advantage from the treatment of her lung disease than he hoped. Circumstances won out. Her last letter to Cabot arrived from southern California, where she had moved for a more favorable climate and a less contentious job. She thanked Dr. Cabot for his efforts to restore her health through the focused treatment of the disease. But she conceded that her original doctor at her home institution had "proved to be right in his estimate of how little I could endure." She was substantially better after a change in circumstances and climate. Medical treatment for bronchitis was powerless, however, in the face of the force of circumstances.[91]

The attempt to dissect out in a fair and disinterested manner the elements of physical disease from a complex of personal circumstances and suffering could prove quite challenging. Writing back about a visit in 1908, a young woman explained to Cabot a year later that she was taking the "opportunity to tell you what happened after I saw you about myself." In the first line of her medical chart from a year earlier, Cabot had singled out headaches as her original problem. Many distinct physical problems could cause headaches, according to Cabot's general reasoning. In an entry on headache in his 1912 textbook, he listed among the possible culprit diseases nephritis, meningitis, syphilis, and brain tumor. She had none of these, according to his initial evaluation. "Always had headaches perhaps dependent on how things go," the chart read, explaining that "whatever goes wrong gives headache." These headaches were a result at least in part of this patient's personal response to her circumstances and so pointed to a nervous condition. Her diagnosis on the visit in 1908 was neurasthenia.[92]

In her letter, this young woman offers an ambivalent response to Cabot's diagnosis and care. The letter sounds initially like the reply of a grateful patient, but it leaves some room for doubt about the benefits of her medical attention. "You may have forgotten," she starts off tentatively, "that I came to you thinking I might have something wrong with me because of my head doing such queer things. From the minute I stepped out of your office until now I haven't worried about it, and I have felt alright excepting three or four times. Isn't that wonderful?" She begins by recalling her fear about something wrong with her head, a fear that she, at least, had not forgotten. Yet she does not explain whether Cabot's diagnosis of neurasthenia dispelled her fears or not. Was neurasthenia "something wrong" or not? Apparently, she had also continued to suffer from the very same troubles, only three or four times over the year, but left unstated whether this marked an improvement or not. She proudly reported that she had not worried about it, at least "until now." But did this report then leave open the question whether "now" meant her fears had just returned? Her closing question to Cabot about whether her condition really was wonderful turns out to seem well justified. The letter conveyed an ambiguous gratitude to the treating doctor who had postulated certain critical distinctions between her headaches, the difficult circumstances of her life, and "something wrong with her."[93]

The characteristics that defined neurasthenia, and later psychoneurosis, in Cabot's practice became increasingly contentious, forcing a distinction between the reality of bodily experiences in nervousness and the evidence of related physical abnormalities. In creating categories of natural disease that were sharply defined and seemingly impersonal, physicians like Cabot found that they had to account for a considerable residuum of personal response both in their patients and in themselves. The claim of medical objectivity met with fundamental difficulties, for example, in assessing the relevance of a patient's Jewishness. Could a doctor claim to be able to reliably evaluate the patient who came to the office suffering from what Cabot described as "Psychic pain—sight of MD often causes"? The physician's claim of objectivity might seem mere inflexibility; and dispassionate regard could appear to be an uncaring response. In managing physically based diseases, Cabot and his peers had begun to back away from sources of personal authority that relied explicitly on traditional hierarchies of race, social class, or gender. For nervous disease, in contrast, the illness became tightly linked to the personality and social identity of a patient, making both nervousness inherently personal and the personal world more openly contentious from the clinical perspective.[94]

CHAPTER 6

MEDICAL CARE FOR THE DYING,

IN PRINCIPLE AND IN FACT

Although Richard Cabot was generally scrupulous in his efforts at public self-disclosure, he kept one episode in his life hidden until his final years; and when he did at last reveal the role that he had played in the death of his brother Ted, he did so under circumstances nearly as remarkable as the event itself. On August 21, 1936, Richard Cabot met in his home with Ada McCormick to discuss the subject. McCormick had been working with him to organize his personal and professional papers in preparation for a biography that she planned and began, but never finished. A transcription of that afternoon's conversation remains with McCormick's research biographical materials, filed among the 255 boxes of papers in the Cabot archives. In preparation for this interview, Cabot went through a collection of his personal letters relating to Ted's death decades earlier and made special arrangements with McCormick to set down his memories of the event. According to McCormick's notes from that afternoon, Cabot told her that she was going to become the only person beside his wife who knew exactly what had happened in November 1893, when Ted died at home under Cabot's care. McCormick, at the end of the interview, appended a comment of personal reflection. After she had heard the full story, she said, she found herself in "a sense of unbelieving shock as if it weren't so." Perhaps she wondered what it was like, as Cabot put it in his own words that afternoon, to "kill the person that you love the best."[1]

The Death of Ted Cabot

In the spring of 1893, Ted was still a young man, but already he had entered the last stages of a struggle against severe diabetes. His brother Richard Cabot had graduated from medical school in the previous year and was finishing twelve months of subsequent training as a physician in residence at the Massachusetts General Hospital. As Ted slipped deeper into his illness that year, Cabot began to take a role in managing his older brother's treatments and eventually became one of his main caretakers. The boys' parents, Elizabeth and Elliott, encouraged their medically trained younger son to take part in advising about his care. In April 1893, Elliott Cabot wrote to Richard, inquiring about his thoughts on Ted's treatments. Elliott mentioned certain "trials" of therapy that another doctor had been promoting and asked "whether it's worth while for you to attempt to give a more particular attention to the experiments which Ted is inclined to try." The family felt uncomfortable at times with other physicians, his father explained in the letter. "It would be a great comfort to us & I think to him," he continued, "to know that some competent person was looking after these things," meaning Richard, the newly minted physician in the family.[2] The invitation met with a dutiful reply. Cabot began to consult in detail, in part through correspondence with his mother, about his brother's medical care; and within a few months, he found himself drawn to Ted's bedside as his illness progressed toward its conclusion.

Throughout his life, Cabot remembered his older brother as his idol. Ted was seven years his senior and held his brother fascinated with philosophical conversations and speculation. A year before Ted's death, Cabot wrote to a relative reminiscing about a night when he had sat up late with him, listening to his thoughts on everything "from God to your watch chain, from a wheelbarrow to a star." Although Cabot aspired to be a philosopher himself, he described his appreciation of his brother's skill as the experience of a spectator on the sidelines. He suggested later that Ted was an inspiration to his own efforts at philosophy. "To watch Ted think," he recounted in one note, "is as superior in interest to the finished thought as watching a man improvise is superior to reading the music." Cabot characterized himself as an adoring audience. Watching his brother think, he continued, "is like being present—it is being present—at the creation of the world." Cabot further characterized his experience as an onlooker to his brother's abilities as a reflection of his own relative inadequacy. "It does no end of good to see Ted," he wrote to his relative, "because he outstrips me in every single direction, not merely in certain particular ones." Ted's philosophizing, he noted, "shows a grasp of the whole subject that I never got in my six courses in Phil[osophy]." His admiration for his older brother affected Cabot all his life. He kept for a long while a personal reading journal on philosophy interleaved in the pages of one of Ted's old reading journals that still bore Ted's name on the cover.[3] For many years after his brother's death, he marked the anniversary of the event as a special day for commemoration in the family.[4]

The growing requirements in the summer of 1893 for his beloved brother's medical care were a weighty charge for Cabot. He took it up with great diligence and attention. He had by the summer completed his residence as the house physician at Massachusetts General Hospital and had secured a spot there for the following year as the Dalton Scholar, with his time to be given over mainly to research on blood diseases. During the course of the summer, Cabot received a series of letters from his mother consulting about Ted's situation, and wrote back with his recommendations. He outlined treatments for skin infections and took a hand in the treatment of the diabetes. In the context of medical care in the 1890s, scrutiny of the diet and attention to the relative balance of the proteins, fats, and sugars were the mainstay in the treatment of diabetes. Cabot wrote his mother weighing the various benefits of skim or whole milk for Ted and explaining the advantages of "plain oatmeal" over the oatmeal gruel that they had inquired about. Such advice was part of conventional medical attention to the disease.[5] Cabot was providing his brother with the best that science had to offer for the treatment of diabetes.

But medical care was known to have a limited effect on the course of the kind of severe diabetes that afflicted Ted. Alongside the routine advice about Ted's treatment in Cabot's correspondence an ominous theme emerged. Their mother had apparently been asking to know what lay ahead for her eldest son. Cabot was pessimistic about the future. By the end of his year in residence at the hospital, he had cared for many patients with diabetes who would have been quite sick. Diabetes was an easily recognized, well-defined, and common condition at Massachusetts General Hospital in the day. The outcome for patients who, like Ted, had the overt manifestations of the disease, accompanied by a chemical imbalance known as ketosis, was almost invariably a gradual wasting course ending in coma or infection and death. This experience would likely have been fresh with Cabot when he wrote to his mother about Ted's prognosis: "All he can do is to make the process go slow & I believe it will be fearfully slow. He will not have any sudden coma, but will gradually die of exhaustion or intercurrent disease." Having given this news, he apologized that he "wrote this so cold-bloodedly." He only hoped, he explained, to give an accurate picture of what they all faced in the twilight of Ted's illness.[6]

At the end of October 1893, Cabot wrote to his soon-to-be wife, Ella Lyman, that "Ted is going down hill very steadily," adding that he was planning to travel to his parents' house to be at the bedside. "I can only be thankful for it," he added, "as long as he doesn't suffer—the quicker the better now." The end that he prognosticated to his mother seemed to be approaching rapidly. Cabot wrote twice more to Ella from his parents' house on November 3 and 4. He was sitting up nights by his brother's bedside tending to him in his difficult moments. As he frequently did in his correspondence, Cabot figured up his worth in these tasks relative to those around him. He judged himself generously on this occasion, noting that Ted "has some glimmering of my love for him this time and I think I can nurse him in many details better than the others." But the work was difficult. Cabot later recalled that it was "horrible to see Ted suffer—the struggles and convulsions." He was dosing his brother with a sedative medication, sulphonol, as he reported in his letters to Ella, and treating him with morphine to ease the pain.[7] The Cabots were rapidly losing their much beloved son and brother.

A Well-Kept Secret

Here the remaining collection of archived letters from 1893 stops, leaving only Cabot's account in the interview with McCormick in 1936. A suspicious discrepancy is evident between the archived correspondence and the account in the interview. In his discussion with McCormick, Cabot referred to his letters from November 1893, which he seems to have been reviewing prior to the interview, quoting from his descriptions of Ted's death to Ella Lyman. Letters to Ella are filed with the transcript of the interview and in the main collections of personal correspondence, but they continue only up through November 4. Ted died on November 10, 1893. The collected letters to Ella pick up again later, but no letters of any kind about Ted's death remain with these files. This gap corresponds to a strikingly similar gap in the records of McCormick's lengthy interview with another of the Cabot brothers, Hugh, who won a local reputation for his readiness to expose or debunk his older brother, Richard. Page 11 of McCormick's interview with Hugh, given later in 1938, records Hugh's comments about Cabot's personal support for euthanasia: medical intervention to end the life of a patient. Page 13 then continues with Hugh's general thoughts on the issue of euthanasia. There is a distinct gap in the interview with page 12 missing from these records. This page very likely contained Hugh's reference to his brother's personal experience with euthanasia, and perhaps a discussion of the events surrounding Ted's death. Hugh referred obliquely to this fact in a letter to McCormick sent several years after the interview.[8] The missing letters that Cabot reported to contain his account of Ted's death, and the parallel missing sections of McCormick's interview with Hugh on the same topic, suggest a later effort to separate out and perhaps to censor material related to this strange episode in Richard Cabot's early medical career.

If McCormick or someone else attempted to remove or conceal the documentary evidence concerning Ted's death, Cabot himself went to great lengths to ensure that the interview that he gave on this subject in his last years would be both credible and available to posterity. The typed transcript of this crucial interview has additional notes in Cabot's handwriting in the margins, clarifying points that were unclear in the transcript or that McCormick perhaps hesitated to set down too boldly. In addition, Cabot added to the transcript a remarkable note verifying his intentions for the interview. Separately attached to the transcript is a page bearing Cabot's signature and the date followed by the awkwardly hand-printed signatures of four children, who "witnessed Uncle Richard's signature in the dining room, not knowing however what the permission to print was about." The note further explained that the children "were the only people handy."[9] Cabot apparently wanted to confirm that he wished this material to be published or made public after his death—as he indicated in a note in the margin—lest others repress this story in the belief that he wanted it to remain secret. The children were made the naïve witnesses to his intent.

Cabot demonstrated a strong desire to have his story heard in full; so we can oblige by considering the account as he stated it that day, although the details are harrowing. His carefully protected interview with McCormick describes Ted's last moments, shedding an oblique light on Cabot's early experience as a physician to his family. Richard Cabot put an end to his brother's life, using chloroform to anesthetize and finally suffocate him. There were two features of the story that may have triggered Ada McCormick's "sense of unbelieving shock," when she heard the account on that August day in 1936 in the study of the Cabot house. First, Cabot related how he had to persuade his parents to allow him to carry out this act. His mother, Elizabeth, he told McCormick, had been the easier to convince. She had been working alongside Richard, nursing Ted through these final days, and had witnessed his suffering. She apparently shared her younger son's views that a quicker end would be more merciful. His father, Elliot, however, had removed himself from the sick room and so, Cabot reasoned later, was slower to acquiesce to the merciful use of chloroform. He recalled going "to father's study several times within an hour pleading for his permission." He argued that an assisted death was the proper thing for Ted. His father finally relented. Having secured this consent, Cabot did not raise the question with Ted. He compared it later for McCormick to a situation in which a physician's patient is requesting an end his suffering: "It's very easy when someone is begging you to kill them to consent & do it but to propose it to someone would be beyond me," Cabot said. Ted had not, apparently, been seeking to be killed, nor did Richard seek his brother's agreement in this matter. The beloved eldest son did, however, come to understand clearly his younger brother's intentions. The manner of this revelation was the second element in the story that seemed likely to shock.

Sitting in his study that afternoon in 1936, Cabot described to McCormick the difficult process of administering a lethal dose of chloroform to his dying brother. It was an awkward and ugly event in his memory of it. He recalled that after he thought that Ted was dead, he began to speak about him in the past tense, when suddenly a whispered voice rose up to him: "Richard gave up too soon—I wasn't dead," his brother said. Even reported from the distance of forty intervening years, the words were chilling. They seemed not so much to be directed at Cabot as to be about him, a gentle rebuke from the elder to the younger brother. Cabot told McCormick that it had not been easy to finish what he had begun and, in his words, "to crush out his struggling life with chloroform." Finally, his brother's misery was ended. Cabot had been equal to the task after all, although he agonized to McCormick on that August day that one had to consider "how much you are doing it for the other person & how much to relieve your suffering at their suffering."[10] The lingering memory of that day stayed with Cabot throughout his life, pressing him perhaps to make such careful preparations for the interview with McCormick that at last enabled him to set it all down in its sad detail.

Ted's death was certainly a tremendous burden for Cabot to assume, especially coming so early in his professional life. As a physician and later a professor of ethics, Cabot published almost nothing about the doctor's responsibilities to patients at the end of life. Yet he tacitly committed himself to a controversial position early in his career by extending his elder brother's medical care to the point of euthanasia. Cabot spent a lifetime with this closely guarded family secret and then made elaborate arrangements to reveal it near the end of his own life. His ambivalence about Ted's death reflected a divide also among other physicians about the responsibilities that attached to care for dying patients. Ancient Western medical traditions had sometimes taught that the physician should retreat from a patient's case as death approached. By Cabot's day, however, other teachings had gained sway that suggested that physicians did have a responsibility that extended up to and through the moment of a patient's death. These responsibilities, however, remained at best incompletely defined, and they were certainly daunting in the range of their implications.

Euthanasia

The views of Cabot's colleagues on euthanasia resembled those on placebos in at least one minor respect. Like placebo treatment, active euthanasia of the kind Cabot had performed was easier to support tacitly than it was to defend publicly—or do. In their interviews with McCormick in the 1930s, both Richard and Hugh Cabot separately said—presumably in relation to Ted's death—that doctors not infrequently practiced euthanasia but rarely discussed it. McCormick raised the question first with Richard. "And yet lots of doctors must do this?" she inquired at the end of their discussion. "I imagine so," Cabot replied, adding "but see they can't ever tell so we don't know." Hugh later confirmed this in his usual blunt manner, telling McCormick in 1938, "I don't suppose there's any doctor who hasn't done it and kept still about it because everyone would yell."[11] As Hugh noted, the professional medical literature of the day tended to raise the question of active euthanasia for dying patients only sporadically and with evident caution. This ambivalence about euthanasia was understandable given the concerns that the practice raised—both over the question of unrelieved suffering at death and over the control of the practice itself. But such ambivalence pointed also to a deeper dilemma gaining force in the medical practice of the early twentieth century, which was equally evident in the records of Cabot's private clinic. What role would doctors properly assume in caring for their patients as death inevitably approached? Among the many segments of the medical profession in the period, the responses were heterogeneous.[12] But a handful of central themes emerge from the literature that defined the problem for physicians in the twentieth century.

In the 1930s, Richard Cabot and his brother Hugh used the word "euthanasia" to mean steps taken to relieve a dying patient's suffering by hastening death, but they were using a medical term that had changed profoundly in its significance during their lifetime. Euthanasia in their sense represented a recent, self-conscious addition to debate at the end of the nineteenth century about the proper role of the physician at the deathbed. Discussion about medical intervention to end the life of a dying person first appeared in the Anglo-American medical literature in the mid nineteenth century, and it only became routinely referred to as euthanasia later during Cabot's youth. The word "euthanasia" in its older, traditional sense meant something else.[13]

We should start by acknowledging that the evidence for this change comes largely from medical rhetoric rather than from medical practice. A good part of what we know about euthanasia derives from the deontological literature, urging on physicians their proper duties or restraining them from improper acts. This literature suffers, of course, from the well-known gap between what people say they *should* do and what they actually do. The practice of medical care for dying patients no doubt differed from what was advised, even among the small minority of practitioners who read the deontological literature. These debates, when they did air, were charged with weighty implications for the doctor's reputation and livelihood. Positions on euthanasia and the proper role of the physician at the deathbed were often articulated most clearly by dissenters who sought a pithy summary of the practices of their opponents in order to show where they went wrong. Essayists in the medical literature typically aimed to persuade rather than to describe.

A reassuringly tidy chronology emerges nonetheless from examination of the word "euthanasia" in the slow-changing and conservative medium of medical dictionaries. In the late eighteenth and early nineteenth centuries, English-language medical dictionaries tended to lack an entry for "euthanasia," although the term had already found its way into the general medical literature of the day.[14] By the mid nineteenth century, however, the transliterated Greek word "euthanasia," defined as meaning "a good death," had established itself permanently in medical dictionaries.[15] Mercy killing as a definition of euthanasia, in the sense that Hugh and Richard Cabot used it in the 1930s, had to wait until the turn of the century to make its way into medical dictionaries, and then it continued to share space with the older meaning of "a peaceful or easy death." Later usage in the twentieth century continued to cite the less controversial, gentler meaning of a good death first, giving mercy killing as a second definition.[16] By the 1970s, however, with the vigor of debate over medical euthanasia, the latter use crowded out all but a brief ceremonial reference to the earlier sense of the word. So the *Dictionary of Medical Ethics* reported in 1977 that "the word is now generally restricted to mean 'mercy killing', the administration of a drug deliberately and specifically to accelerate death in order to terminate suffering."[17] Euthanasia had completed a transformation in its reference: from the experiences of a dying person to the actions of a presiding physician.

This change in terminology hinted at deeper currents in nineteenth-century medical care for patients near the end of life. Essayists of the time mused on a variety of possible roles for the physician. Although doctors had once tended to back away from direct involvement with patients who were judged to be dying, they might need to stay. The more ancient advice to avoid involvement appealed to practitioners who were anxious to sidestep the charge that they had promised a desperate family or patient more than medicine could provide, and it allowed them to make way for final rites that seemed appropriate to the family, religious officers, and the law. But physicians might still preside in a ceremonial way over the deathbed, witness the event, validate its gravity, and assist in the small humane ways that physicians often recommended to each other in the absence of bold therapeutic strategies. "The physician should be the minister of hope and comfort to the sick, that by such cordials to the drooping spirit, he may soothe the bed of death, revive expiring life, and counteract the depressing influence of those maladies which often disturb the tranquility of the most resigned in their last moments," the American Medical Association's first code of ethics advised in 1847.[18] The code continued by recommending that the physician not abandon a patient just because he was deemed incurable. Physicians should become comforters and supporters of the dying patient and the family. The use of direct euthanasia seems a long way off.

By Cabot's day, a medical presence at the deathbed was uncontroversial, and physicians advocated increasingly active roles in caring for dying patients. In the later nineteenth century, a pair of sharply divergent approaches to the care of the dying emerged in the American medical literature: attention to dying, or attention only to final disease. On one side, physicians acknowledged an explicit duty to care for dying patients. They tended to advise detailed treatment of the symptoms of the dying to relieve the associated suffering. Common problems like pain, breathlessness, chills, and anxiety might be addressed with medications and strategies, independent of the underlying disease. But this approach receded to a minor theme in the literature. A newer model was gaining sway, in which obligations to treat diseases changed very little in the care of the dying. With a growing range of disease-oriented medications and procedures, physicians might continue to treat specific diseases up until the time of death. Relief from suffering might be achieved indirectly through a continued attack on the patient's underlying pathology. Even if it were possible only to extend life marginally, continued treatment of the final disease might still assist the dying patient by maintaining hope and indirectly palliating disease-related symptoms. This latter approach also fitted in neatly with the traditional goal of medicine to work toward amelioration and cure, however unlikely. Also since the final illness might well prove to be only the penultimate illness, with advances in medical therapeutics and technology, this approach may also have seemed more prudent. Physicians who endorsed one or the other approach differed starkly in their portrayal of the doctor at the deathbed. But both divergent views related at the extremes to the contentious possibility of active euthanasia, increasingly acknowledged by American physicians by the turn of the century.[19]

The Physician's Role at the End of Life

One passionate proponent of the physician's explicit duty to acknowledge and treat the onset of dying was Cabot's close associate and colleague Alfred Worcester. In his writings, Worcester espoused a tradition of medical advice that focused on a patient's symptoms and personal responses rather than on diagnosis or specific disease processes, which were regarded as unimportant in the care of dying patients. William Munk, a member of the British Royal College of Physicians, offered a concise and lucid set of such recommendations in his 1887 book *Euthanasia, or, Medical Treatment in Aid of an Easy Death*, which Worcester cited.[20] For Munk, euthanasia did not imply active measures to end life but referred instead to the peaceful death that might be aided by a physician who was properly oriented and prepared. Treatment of the dying should be guided above all by the immediate needs and requests of the patient. Munk outlined in some detail, for example, the use of various forms of alcohol to treat the symptoms of dying, demonstrating a liberal solicitousness in his advice. Weighing the differing virtues of sherry, port, champagne, and brandy as medications, he concluded, "The wish of the patient for any particular form of stimulant is almost always a correct indication for its use." Munk rarely mentioned specific diseases or disease processes. It was the patient's reported perceptions that mattered most. He kept a similarly tight focus on the patient's suffering, recommending "the relief of pain in the dying wherever it may be situated."[21] He eschewed the doctor's usual question about where it hurt or why, asking only how much it hurt, and how much better it got. Opium, the derivative morphine, chloroform, and other anodynes would need to be generously applied. The principal guidance to the prescribing physician here was the patient's distress.

Forty years later, in 1929, Worcester's advice on "the care of the dying" had changed very little in substance or goals from what Munk advocated. The core practical elements remained the same. Treatment for the physical discomforts that commonly accompanied dying, like dry mouth, breathlessness, and pain, was paramount, with the liberal use of narcotic medications like morphine as a mainstay. Worcester reported an enthusiasm for the relief of pain that only amplified Munk's. "Opiates are indispensable," he advised: "If morphine fails to give comfort, a hundred to one it is either because too small doses have been given or because it has not been successfully introduced into the progressively enfeebling circulation" of the dying patient. Worcester dwelt rather less than Munk on the virtues of alcohol, however, recommending only wine, and then with water to dilute it.[22] But the advice of Munk and Worcester was nearly interchangeable.

This minor tradition of practical advice on the care of dying patients remained remarkably stable in twentieth-century medical literature. Advocates of the approach continued to offer injunctions similar to those issued by Worcester and Munk throughout the twentieth century. In part, we must imagine that this stability stemmed from relative neglect. Teaching about the general care of dying patients in the United States received only marginal attention in Cabot's day, and it only slowly developed its own orthodoxy in the late twentieth century, with the creation of hospices and the emergence of formal status for a field known as palliative care.[23] Medical advice in this tradition, when it did surface, however, remained consistent. A half century after Worcester, Dame Cicely Saunders, the British nurse and physician who became the international leader of the hospice movement in the 1960s and 1970s, preached a similarly enthusiastic line about symptomatic treatment for the dying and about the value of morphine. Anecdotal teachings about specific pharmacological interventions at the end of life were repeated without much elaboration. Like Worcester, Saunders recommended a mixture of hyoscine and morphine to sooth a patient's labored "death rattle."[24] The greatest area of change by Saunder's day was in recommendations about the grief, anxiety, fear, and despair that attended dying, which by the later twentieth century derived counsel from detailed psychological theories, in place of what Worcester called "an appreciation of the dying patient's personality."[25] There was otherwise little in the content of Saunder's teachings that was new in itself, though she was an inspired teacher and advocate in her own right.

In contrast to these simple symptomatic approaches to the care of dying patients, the dominant medical alternatives gained considerably in complexity and scope over the century. Treatment of the final disease became a rarely questioned extension of the management of disease more generally. Worcester in his short essay on "The Care of the Dying" in 1929 did not distinguish dying patients in any way according to their diagnoses or diseases, following the format set out by Munk. Advocates of this general medical strategy for the care of dying patients had to contend with the increasingly persuasive suggestion that disease itself should remain the doctor's principal concern, even in the face of impending death. By attending directly to a life-threatening disease, physicians could do what they did best and still palliate the dying person's suffering indirectly through their prescribed treatments.

The Treatment of the Final Disease

The "concealment of death by illness" is the apt phrase of Philippe Ariès, who has written on changing European-American conceptions of death over the past half-millennium. In the past few centuries, Ariès contends, a growing preoccupation with fatal illness blunted attention to the process of dying and the event of death. Death came to be acknowledged only once it had passed, while attention to proper deathbed preparations was slowly obscured by the treatment of the final illness. Ariès identifies a shift in focus from death to fatal illness as affecting Western culture widely over a vast swath of time. With his expertise and interest in European Catholic traditions, Ariès finds his evidence in examples like the gradual replacement of priests by doctors as the chosen attendants at the deathbed and in the incremental postponement of various religious observances until after death; but he also allows correctly that any fundamental change in Western attitudes toward death must certainly have been partial and heterogeneous.[26]

Ariès's narrative of essential, overarching change is difficult to apply in any detail to medical writings. But his key formulation does help to sharpen distinctions regarding the physician's changing role at the deathbed. Medicine did seem to assist in moving the final illness to the forefront in attention to the dying. Developments in European and American medicine in the nineteenth century lent support to a strategy of attending closely to disease processes during dying and gave ample reason to distinguish these processes from some general, undifferentiated process of dying. Attention to a final disease defined a set of obligations for the physician that merged seamlessly into the increasingly intense monitoring and management of disease mechanisms. Physicians over the nineteenth century dramatically expanded the manifold duties that they attached to the treatment of disease and made few explicit exceptions for diseases that were anticipated to end in death. What relief doctors did provide for the dying might properly emerge from the treatment of a final disease. Shifting the focus from dying to disease also provided a possible distraction from the disturbing fact of death in difficult moments—not only for physicians but for family and patient as well. Whatever the breadth and source of a cultural anxiety to "conceal death"—as Ariès would have it—medical expertise became one available means for such concealment.

Although this medical rationale held a paradox at its core, it remained a compelling dogma. To fight against fatal disease meant inevitably to lose; and if this battle were the sole medical service to a dying patient, physicians might count it a meager contribution. Perhaps with the recognition that death was near, doctors would redirect their skills away from fighting disease and toward easing any suffering that accompanied the patient's departure from the world, as Worcester and Munk would have it. But the ideal of a noble medical struggle, no matter how futile, could amass powerful cultural significance and be variously deployed. Writing in 1948, in the aftermath of Britain's punishing endurance through World War II, one English physician hailed the doctor's role at the deathbed as "a most satisfying reward—the culmination often of a long battle with death that patient and doctor have fought together with courage and determination; the acknowledgement that although the last enemy has won the patient has still confidence in his physician and that the human spirit can triumph even in the hour of defeat."[27] If struggle in the hour of death was a triumph, was palliation then a less noble pursuit, an unheroic resignation to defeat? It was a rhetoric that supported a focus on final therapeutic measures.

The emerging twentieth-century strategy of treating only the final disease had tangled roots. Earlier practices of diagnosis and medical classification that supported attention to dying as a general medical phenomenon worthy of separate consideration were gradually receding in the nineteenth century. The idea that extreme age was a cause of death lost much credibility in the nineteenth century, as the process of aging itself resolved, at least partially, into the cumulative effects of specific diseases. The historical sociologist Hans-Joachim von Kondratowitz has shown that late eighteenth-century German physicians who defined aging as a nonspecific and irremediable process leading to death were challenged by their colleagues. The struggle remained mainly within the bounds of medicine. *Marasmus senilis,* a generalized wasting condition of aging, had long been considered, as one eighteenth-century physician put it, "the desiccation and wasting away of the old, to which they eventually abandon their spirit, because their juices can no longer circulate; therefore remedies are completely useless and death is unavoidable."[28] By the mid nineteenth century, however, German physicians argued that attributing death to old age had become an inappropriate shelter for physicians seeking to escape responsibility in such cases. The following passage from a mid-century German scholar quoted by von Kondratowitz is worth considering in detail, since it connects several arguments against the acceptance of the dying process as a unified medical concern:

> The value of pathological-anatomical experiences for the study of the illnesses of extreme old age is even more evident when one considers the paralyzing influence of the customary term of marasmus senilis on the less zealous of our fellow craftsmen in the making of a diagnosis. The public, otherwise so incredulous, listens credulously to the doctor when he utters the portentous words "old-age infirmity." Then there is nothing more to be done, and if the sick person dies, this is perhaps the only case where a higher hand has ordained it and no blame is attached to the doctor.[29]

Physicians needed to be encouraged or pressured to identify specific diseases near the end of life, so as not to shirk their fundamental responsibilities to their patients—and to their science.

Heirs to this established literature and ethic, American physicians in the late nineteenth century formulated an increasingly urgent case for the need to attack disease and its physiological derangements, even in the face of death. The descriptions of care for dying patients in the literature show evidence of a growing concern to forestall death by shoring up the failing body. In 1890s, as Cabot was pursuing his medical education and training, American medical journals frequently carried detailed accounts of individual cases. Very few of these accounts ended with the death of a patient, as might be expected, but when they did conclude in death, the report was typically terse and impersonal: for example, "died in uremic convulsions, two months later," or "gradually sank and died," or "died in about one minute without any struggle."[30] Occasionally in such accounts, physicians also related the details of how they doctored people through their final hours. Thus, for example, Dr. J. Holyoke Nichols writing about a patient slowly dying from smallpox, described how he used salol to treat bladder disturbances, enemas to loosen the bowels, chlorate of potash for "nasopharyngeal disturbance," and bismuth powder on the sheets to absorb skin discharges.[31] But other physicians in the same editions of these journals presented their work in a more dramatic light. Surgeons seemed particularly committed to this tack. In 1896, a Boston obstetrical surgeon described injecting a "pint of salt water" into a fatally hemorrhaging patient "on the bare chance that stimulation might restore her to a condition that would warrant an attempt at surgical arrest of the hemorrhage." Another obstetrician noted that repeated administration of chloral hydrate and chloroform could not arrest his patient's terrible seizures, which led steadily to coma and then to death. Such accounts highlighted the regret of another surgeon who speculated in reporting on his patient's death that "had I boldly opened the abdomen . . . I believe a life would have been saved."[32] Perhaps there was always one final medical measure that might turn back the tide.

This heroic impulse drew on established nineteenth-century ideals, while incorporating new confidence in the powers of medical treatment. Such accounts implied a general reconfiguration of the twentieth-century physician's role at the deathbed. Anecdotes about the physician's final stand against fatal disease predated Cabot. But by the mid twentieth century, the ideal had become dogma, wedded to the striking claim that death might in fact be understood simply as a failure of medical therapy. Ernest Daland, writing on "Treatment of the Patient with Advanced Cancer" in 1948, proposed that "every death from cancer is indicative of failure somewhere along the line." He too invoked the dignity of final struggle against disease, suggesting that "the patient with cancer never gives up hope and will fight as long as his physician stands by." All the physician's measures had to be interpreted in light of this battle against disease, in Daland's telling. So, even the doctor's attempt to palliate was really better thought of in other terms: "not infrequently treatment intended to be palliative may be curative," he claimed.[33] Hope reigned supreme in therapeutics, seeming to allow little room for a valid acknowledgement of the general problems of dying itself.

Cabot's peers in the early twentieth century voiced as a theoretical possibility the repeated, indefinite postponement of death by treatment of disease.[34] By the last third of the century, physicians had secured this tenet as a working convention in certain high-stakes areas of practice. Renée Fox and Judith Swazey have documented the notion that death was equated with a failure of therapy among physicians working in transplant surgery, and later hemodialysis, in the 1960s and 1970s. In Fox and Swazey's estimation, at least among the highly specialized practitioners whom they studied, medical training and experience combined to inspire physicians to act "as if death ideally should never occur and they should always be able to prevent it."[35] The struggle against final disease could always be won episodically, even if it only bought a brief respite from what would increasingly be understood as a tacit failure.

The Problem of Therapeutic Excess

In the later twentieth century, the palliative methods that Worcester taught anecdotally became formalized and gained a new rationale. Palliative care, or general symptomatic treatment near the end of life, became a means to reinforce therapeutic restraint. Physicians writing on palliative care began to offer their craft as an explicit solution to the problem of excessive medical treatment for dying patients. Worcester in 1929 did not represent the symptomatic care of the dying as a response to overly aggressive attempts at cure. He did caution that such attempts might be counterproductive. "Modern methods of resuscitation," Worcester advised in the 1940 edition of his essay, "which of course are obligatory where valuable lives might thus be saved are most decidedly out of place where . . . the body's usefulness has ended."[36] By the last part of the century, however, this line of argument had been extended into a programmatic defense of the growing field of palliative medicine. In an effort to secure support, advocates of palliative care co-opted the language of heroic therapeutics. So a founding advocate, Cicely Saunders, argued that good palliative care meant exactly that "everything possible is being done."[37] If "everything" included only efforts at the treatment of disease, and not efforts to quell suffering, then the toll on patients could be very high indeed, by this accounting. Saunders argued that such rationales in palliative care had become necessary, since few people seemed likely to relish the idea that *not* everything was being done for a dying relative or patient.

These programmatic defenses of palliative care amplified earlier, minor strains of resistance to desperate, therapeutic attempts for the dying patient. By the mid twentieth century, physicians writing in the American medical literature were demonstrating a practical concern with the merciful restraint of such interventions. It seemed increasingly necessary to make "a plea to the physician to refrain from unnecessary heroics," in the words of one Chicago surgeon addressing his professional society in 1957, who cited his experience with numerous patients who had undergone terribly destructive and futile surgeries for what were clearly incurable heart diseases and cancers. He offered the understated observation that he and his colleagues were liable to find themselves "going through the motions in order to kid ourselves and the patient's relatives that everything possible is being done," for the patient. Mid-century physicians had begun to see that as the range of "everything possible" in medical therapy grew ever larger, so too did the risks hazarded by those "going through the motions." The physicians of the mid twentieth century were living, as this Chicago surgeon put it, "in the golden age of medical miracles," yet these miraculous cures, when they failed to be miraculous, might exact a devastating toll. Earlier physicians had acknowledged the hypothetical risk that "modern methods of resuscitation . . . would only renew the patient's sufferings," as Worcester put it. By the middle of the century, physicians like this Chicago surgeon had logged considerable personal experience with the precise forms of suffering that could be exacted through prolonged efforts at heroic treatment of a final disease, as this physician catalogued in affecting detail.[38]

Therapeutic excess at the end of life was not the same problem for patients and physicians in Cabot's day that it would be later. Yet a similar ambivalence about the role of disease-oriented therapy for dying patients appears in correspondence between Cabot and his contemporaries. The limits of medical treatment were often on Cabot's mind, particularly in noting the presence of a fatal disease. He did not present the palliation of symptoms as an explicit means to ward off the lure of excessive treatment, as advocates like Cicely Saunders would later do. Yet this tension between treatment and palliation is evident in his correspondence all the same. He wrote to one physician, for example, to offer some advice regarding an older woman with a cancer that "the case is altogether beyond the reach of surgery, even if she were young enough and vigorous enough to stand so formidable an operation. All that we can do is to make her as comfortable as we can."[39] Comfort was the alternative to treatment. Cabot was not one to mince words when he had such news to deliver. He generally made his point succinctly, writing, for example, to a family member concerning a young lawyer who was his patient: "I fear there is no doubt but that he has pernicious anemia and I cannot hold out any hope of recovery." Cabot seemed to accept that the identification of an untreatable disease like pernicious anemia carried with it the implication of death.[40] Treatment opposed disease, so that the absence of effectively specific treatment meant the triumph of disease.

Even when reliably powerful treatments did not exist, however, Cabot did not abandon the promise of rescue in some cases. He was generally optimistic about the potential of good general hygiene and nurturing to affect a difficult disease like pulmonary tuberculosis, and he was known among his peers for his readiness to encourage an activist approach to care for this disease. Patients also on occasion testified to his ability to inspire their optimism in this respect. One man wrote in 1915 that he was recommending that a friend contact Cabot, since her child was dreadfully sick and the child's "physician offers no hope but I have urged her mother not to give up the fight until we hear from you that there is no help for her. Some years ago you gave me the lift that started me well out of the grip of tuberculosis. Now I am perfectly well and it seems natural to turn to you in a case that seems nearly hopeless."[41] Hope itself seemed a valuable service that the physician could provide. Physicians corresponding with Cabot often spoke of the responsibility to maintain hope. They shared a belief in its value with the families of the patients too. The economy of hope presented a complex set of tasks to physician, patient, and family in the face of impending death.

Extremes of Care

Physicians were often at the center of the fray over euthanasia in public debate during Cabot's lifetime. It remained unclear, however, which medical traditions might lay claim to the practice, and it was widely disowned as inappropriate to any strategy for treating final illnesses. Was euthanasia an extraordinary extension of the effort at palliation, or was it some form of final, heroic intervention against the destructive activities of disease? The staunchest advocates of palliative approaches to the general problems of death typically denied an interest in euthanasia, perhaps because of the potentially controversial similarities in the practices. Munk, in 1887, titled his book *Euthanasia,* in the sense of an easy death, before the meaning of "euthanasia" shifted. He stated simply that measures to provide comfort to the dying need not extend to the point of shortening life. The idea that the measures Munk advocated, like the use of ether, might aid an earlier, easier death had barely found recognition by his time. Worcester at a later date contended more vigorously that the intent of his advice was not to shorten life. He went further than Munk in arguing that the generous use of morphine actually extended the lives of his dying patients. Later advocates in this tradition made the case even that the success of palliative measures should eventually displace any need for active euthanasia, by eliminating suffering without eliminating the patient.[42]

Euthanasia might also appear to be the extreme at which palliation and disease therapy met. In the context of medical records from Cabot's practice, the act of euthanasia, when it appears, seems almost a Pyrrhic victory. The physician's attention was focused on the final disease and efforts to contain it or avert its harsh effects on the patient. Euthanasia seemed a final response to the progress of a devastating disease. Doubtless this impression could misrepresent the physician's intention, perhaps to provide comfort in the final moments of life. Yet, without any acknowledgment of the patient's experience of these measures, the medical chart, with its terse style, offered only a picture of a failing struggle against disease. Euthanasia seemed an ambiguous solution to the problem of medical care at the end of life. The physician arrested the advance of disease finally through the patient's death. When Cabot himself entered the term "euthanasia" in medical charts, it seems like the last of a series of medical interventions of diminishing effectiveness.[43]

Although Cabot had extensive contact with people in the late stages of terminal illness, his contact with individual patients near the time of death was rare. The exceptions are, however, compelling in their details. A middle-age woman who arrived at his office in September 1900 with "much worry" about a lump that she could feel in her right breast (see chapter 3) endured an odyssey of medical care, with Cabot as one of her principal physicians. The lump turned out to be breast cancer, which eventually killed her, and Cabot's records show his close involvement in treating her condition and its complications up to her last day.

During the months after this woman's initial diagnosis of breast cancer, she and her doctors seemed to make progress against the disease. Dr. Charles Porter, who had a private office on Beacon Street near Cabot's, performed the biopsy that identified the lump as cancerous and later operated. Although Cabot's medical records do not identify the nature of the surgery, it was presumably excision of the cancer, since it left a wound in the area of her right breast that took months to heal. The woman subsequently saw Cabot for postoperative care, visiting his office in January and March of 1901. The records do not indicate at this point what either this patient or her physicians concluded about the success of the surgery. Porter and Cabot may well have been aware of the optimistic reports of the surgeon William Halstead at the Johns Hopkins Hospital, who had published an account just a year earlier of the successful surgical cure of breast cancer by aggressive surgical removal of the cancerous mass and the surrounding muscle and tissue.

During the patient's office visit in March, Cabot found that an abscess had formed, which he drained twice over the next few weeks, finally enabling it to close up. Cabot noted her gradual improvement over six successive visits during the summer of 1901. She was able to resume her previous work briefly, and then to enroll in some classes at Harvard College in the fall.

Her condition then took a turn for the worse, however. She began to suffer from back pains, and the chart notes the gradual exacerbation of these through October. By November, the pain was radiating from her back down into both her legs; she had "been very blue for a few days," Cabot noted during this visit. On November 15, 1901, he saw her again and noted tersely, "gait reduced to a miserable shuffle."[44]

It had been a rapid decline; yet there remained things that could be done medically. Although he described his patient as "bed-ridden," Cabot made arrangements for her to travel from her home to Massachusetts General Hospital in Boston, where an x-ray photograph of her lower back was taken in November 1901. The x-ray confirmed the destruction of the bones in the spine just above her pelvis, likely from the extension of the cancer. Cabot advised a "leather jacket" to support her back and to slow the further collapse of the spine.

In January 1902, he took notes from a visit to her bedside at home. She was fearful "in spells," and an accumulation of fluid, presumed to involve her right lung near the site of the original tumor, seemed to be causing her difficulty breathing. (An autopsy after her death would confirm that the cancer had spread to the interior surface of her right chest wall in the space around the lung, explaining the accumulation of fluid.) Seeking to provide some relief from the breathlessness, Cabot "tapped" the fluid, drawing off a portion of it with a large, hollow needle inserted into the space through her chest wall, which gave some relief, presumably as the lung expanded back into the full area of the chest.

Near the end of his patient's life now, Cabot faced rapidly narrowing choices. Indeed, of his next house call, he noted only the date, in early 1902, with a terse, summary conclusion, on a line by itself: "euthanasia." Presumably this indicated medical intervention to shorten her suffering, as Cabot would explain it in his interview with Ada McCormick in 1936.

The records of the case continue briefly with details of the autopsy done the next day by one of Cabot's colleagues, the pathologist James Homer Wright, which showed the rampant spread of the cancer through the patient's internal organs, lower spinal vertebrae, and chest wall. Halted by her death, the cancer was frozen in time under the eye of the pathologist. Surgical treatment had failed to contain it, but it would go no further now.

The Medical Discussion of Death

Although Cabot's chart often seems tightly focused on the treatment of a final disease, his records preserve general observations about the problems of the dying. Physicians writing to Cabot provided detailed accounts of their struggles against disease up through the last moments of a patient's life. In 1913, one doctor laconically related a dramatic series of assessments and interventions he had made for his dying patient. The seeming imminence of death did not excuse discontinuing attention to the details of bodily function. He recited his final medical maneuvers in a letter soon after his patient's death: "The right lung was completely involved and seven hours before death expectoration ceased. The delirium became very troublesome and the heart went to pieces regardless of stimulants. Oxygen gave only temporary relief." In this doctor's account to a medical colleague, the patient was not simply dying but rather failing successively by parts—lungs first, then cognition, and then heart—and failing similarly to respond to treatment.[45]

The letter is a fascinating bricolage of the perspectives of doctor and patient, assembled for the purposes of the narrative. The doctor's assumption of both perspectives was partly a convenience of rhetoric, but it also gave a means of explaining his medical interventions on behalf of a patient who had died. The observation about the problem in the right lung was certainly from the doctor's perspective. But the temporary relief that the oxygen provided must have been the patient's, although his doctor left unclear whether it was a general relief or only a relief of his disintegrating heart. This letter is similarly ambiguous in relating just who it was that found the delirium troublesome. The delirium might have been difficult for this physician to witness and to manage, although he seems also to suggest that it was difficult for his patient to suffer through. The assumption of the patient's perspective sheds light on the hazardous and troubling problems of treating disease at the boundaries of the patient's death. The observation that the heart went to pieces despite stimulants was certainly the testimony of the physician attempting to treat it. But the goals of his treatment are unstated. Was this effort at treating the heart to be understood as a heroic attempt to save the patient or extend his life? Or was it intended instead to provide some relief from the suffering that seemed to attend the anticipated process of dying? Or was this treatment simply the continuation of the doctor's ongoing strategy of fighting the underlying disease under any circumstances? This letter gives the physician's account of a patient's death, but it hints at the presumed responses of the dying patient as one means to explain and justify these seemingly futile medical activities.

The majority of the observations about death in these records are from physicians, who offer accounts focused on the problems of fatal disease rather than the difficulties of dying. Some of the accounts are Cabot's own, but many came from other physicians with whom he had consulted. Patients seem rarely to have considered mortality in their correspondence with Cabot. He was typically treated by his patients as an ally against personal death who had little interest in news of his adversary. Families did write about a death in retrospect, but typically in a cursory fashion, or, as in this example, almost reproachfully: "would state that [my brother] died very suddenly August, 1904, very shortly after his visit to you."[46] Cabot's usual arrangement of visiting people at home only as a one-time consultant meant that he was rarely present at a patient's death. When he acted as a consultant, he generally left the ongoing care to the attending physician. Physicians with whom he consulted did write to inform him about the deaths of the patients he had seen, sometimes in detail. But these medical accounts have a perspective that often obscures or subsumes the experience of patients at the end of life.

The death of a patient often elicited strong reactions in the physicians who reported their experiences. These doctors reported close contacts with the preparations for death by patients and their families. One was present at the signing of a will on the day before a death, and then again the next day when a shaky codicil was added in the last hours.[47] Cabot was called to consult on patients who were near to death on multiple occasions. A typical advice manual from the day suggested that it was prudent to bring in a second opinion on a patient who was in danger of death, and perhaps the attending physicians who called on Cabot's advice were thinking along these lines. Cabot's prognostications of death are a striking theme in several of these letters from attending physicians, a couple of whom seem to have been eager to convey to Cabot that his pronouncements had been eerily accurate. One doctor reported simply that the patient's death was precisely "the result you predicted at the time that you made your visit." Another wrote to say that Cabot's letter following his visit to the patient had "a 'prophetic strain' in it," because it had reached the physician by post the day after the patient's death, carrying the caution that it might arrive too late for this patient.[48]

Disease as a Means to Dying

Richard Cabot died on May 9, 1939, aged seventy-one, nearly five years after the death of his wife of thirty years, Ella Lyman Cabot, on September 24, 1934. Cabot witnessed the approach of both Ella's death and his own self-consciously, recording his observations in personal letters and autobiographical notes. In addition, he archived some of the other doctors' medical records related to these final illnesses.[49] The evidence of impending death in these records was measured out by Cabot and by his doctors in the medical discussion of his and his wife's diseases. But if the disease seems to be a screen in front of the daunting problem of personal death, it serves better at times to reveal the problem than to hide it, showing the shadows cast from behind. Cabot had heart failure, with symptoms that were already evident to him before Ella's death. He spent the final year of his life treating this condition under the care of the renowned cardiologist Paul Dudley White. Five years earlier, Ella had died from a cancer that Cabot himself first diagnosed. He provided her initial medical care before ceding the responsibility in her last days to another colleague, Dr. Roger I.

Richard and Ella Cabot on the porch of their house in Cohasset, Massachusetts, in the summer of 1933, the year before Ella's death. Courtesy of the Harvard University Archives.

Lee, at Massachusetts General Hospital. The records of these terminal illnesses, for husband and for wife, document the profound difficulties of
accommodating the fact of death within medical responsibilities for disease.
The physicians who attended Cabot and his wife never expressly organized
their responsibilities around the problem of death. Yet the management of
the final illness was not impervious to the acknowledgement of death, as the
medical documents superficially suggest. Both Cabot and his wife seem to
have been able to find ways to accommodate the challenge of personal
death, even within a medical framework that ostensibly excluded it.[50]

Cabot's observations on his own final illness nearly swallow up his
reflections on his mortality. In mid January 1939, the condition that Cabot
called his cardiac weakness progressed suddenly, leaving him acutely sick.
He had been suffering for several years from the progressive manifesta-

tions of heart failure and would die of it four months later. Yet the brief notes that he and his attending physician left about the episode in January barely address this possibility. In 1926, Cabot had published a seven-hundred-page book on the diseases of the heart that outlined their classification, their pathology, and their prognosis. Concerning the common condition of cardiac weakness, he noted that "death usually follows within a year from the first serious evidences of stasis [heart failure]." His episode in January certainly qualified as serious evidence of heart failure, and not the first. Yet Cabot and his doctor documented assessments and plans that made no allowance for the prospect of imminent death. The records from the event show instead attentions squarely fixed on the elaborate set of medical treatments necessary for managing the late stages of heart failure. Death remained an unexamined and unacknowledged possibility, although it was the naturally anticipated result of medical strategies that were rapidly running their course.[51]

In January 1939, Paul Dudley White responded to Cabot's suddenly worsened disease. He began treatment and when he had to leave town for the weekend, he left a note dated Friday, January 20, with a series of orders for the physician who was covering for him during his absence. Dr. White, like Dr. Cabot, had also written a very large textbook on heart disease, published in 1931, weighing in a bit over Cabot's book at nearly nine hundred pages. The list of medical orders for his patient's care that Friday afternoon neatly recapitulated the twenty-two pages of the textbook devoted to the treatment of congestive heart failure. He instructed that a preparation of digitalis be given daily to stimulate the heart. Dr. E. F. Bland, the covering physician, would administer a mercury-diuretic medication intravenously on Sunday morning to stimulate the excretion of excess fluid. Any sudden increase in breathlessness would require rapid treatment by Dr. Bland with subcutaneous injections of dilaudid, a synthetic narcotic medication; and if that did not suffice, then supplemental oxygen given through a mask. Dr. White's determining goal behind these treatments remained unstated. His nine-hundred-page textbook contained three pages indexed by the word "death," but these reported only the incidence of death from heart disease, providing no practical advice on how to care for a dying patient. Most of the book was reserved for advice on combating the disease.

Would there be any conflict between the goal of treating Cabot's heart disease and the goal of accommodating some fitting end to his life? Cabot might well have gone on to die that weekend from his heart failure—as he did a few months later in May from its recurrence. Dr. White in his textbook gave a very guarded prognosis for the condition in general, especially when it was accompanied by angina, as in Cabot's case. Yet the orders that he left presumed a living patient, and also a patient who might need to be persuaded to act in a way that would best treat his disease. So the orders for Saturday and Sunday briefly specified how Cabot should eat and drink, what he should be advised to do or not to do, in order to manage his condition. The patient would have a "very light supper. No fluids after supper." He would restrict his movements to his bed and chair only and walk as little as possible. He could work quietly, but only three hours total each day, in stretches of one hour at a time. There could be visitors other than family members, but only two of these each day, and with their stays strictly limited to half an hour each. Cabot's activities were thus tightly constrained, presumably in the interests of managing his disease. The advice seems to have been effective, if Cabot's survival through the weekend can be construed as evidence in its favor. Yet the value of these prescriptions as a way to spend what might amount to the final two days of a life is less certain.

We might count this critique as a limitation of human prescience and wisdom as much as a limitation of medical care, but Dr. White's therapeutic strategy also carried certain ambiguous implications for his authority over the care of his dying patient, Richard Cabot. If the authority of the physician derived from the expert management of disease, how did a physician respond when disease threatened to be unmanageable? Neither White nor his patient would have expected Cabot to live long in his condition. Yet White's most pressing duty remained to contend with the last stages of the disease, presumably in a tacitly shared effort to struggle on indefinitely. The approach left his patient with some difficult, unexplored choices. Cabot would not necessarily have the luxury of dealing with death later. When White prescribed a "light supper" with no fluids afterwards, what if he was indeed specifying Cabot's last supper? How much of the validity and value of a doctor's orders was lost if the promise to control disease was not kept? If Cabot decided that he wanted extra fluids, should he drink them? What exactly was the risk, or, alternately, what was to be gained by careful compliance with White's recommendations? Would straying from the regimen undermine the value of other orders for the weekend? The unstated decision to focus narrowly on the disease in the prescribed regimen kept such questions at bay. It was unclear whether Cabot should tell Dr. White about his decision to adhere to or ignore recommendations. Could Cabot reasonably pick and choose among these orders without forfeiting the attentive care of his prescribing doctor?

On the face of it, Cabot's brief notes on his severe illness in January show a similarly narrow attention to disease. The format of the notes suggests that Cabot intended them as a guide for his physician, derived from his own knowledgeable observations on his medical condition. The notes form a sort of diary of symptoms, setting out the course of his progressively worsening cardiac condition. But symptoms served not only as clues about disease but also as experiences of daily life, bringing to mind a host of memories. Speaking as one physician to another, Cabot summarized his problems in medical terminology, listing "two attacks of acute pulmonary edema," when he was unable to breathe comfortably while lying in bed at night, and certain previous "evening attacks of sweating and substernal [chest] pain." These problems were accompanied also by "rales [sounds from excess fluid] audible in my own chest without a stethoscope," as Cabot put it, casting himself in the unaccustomed role of a patient, without the usual tools of his trade. All the signs, as he organized them, pointed to the late stages of heart failure, but since these signs also constituted evidence that he was dying, thoughts of mortality were perhaps not far removed. Cabot's "notes on . . . cardiac weakness" continued in a dutifully informative manner, but also admitted the poignant reflections of one man about his life. Cabot characteristically organized the list meticulously into six categories of "facts possibly associated." But as he worked through his six categories, he linked his symptoms one by one with memories of past places and special people in his life. He was reviewing his life in the guise of a symptom list. Breathlessness, he said, reminded him of hiking in the Swiss Alps with his wife, Ella, and also of his embarrassment in years past when he had had to keep tennis partners waiting for him when his chest discomforts overwhelmed him and he had to rest. An attack of sweating and pain, listed under the sixth heading, reminded him of an exciting conversation with a close friend, while his inability to lie flat during sleep triggered a set of reflections on the sleeping habits of other friends. The history of his heart condition was also an occasion to reminisce, charting the course of his life against the progression of this disease toward its natural conclusion.

Cabot had witnessed the death of his wife Ella just a few years earlier, sitting at her bedside in Massachusetts General Hospital for her final twelve days. He reflected on his experience in the same series of interviews with Ada McCormick that dealt with Ted's death. Doctors might focus on disease, but they could not help but face the reality of death and be altered by it, he suggested. He described Ella's final moments and recited an injunction that he had heard from his colleague Alfred Worcester. "[I]t was the greatest shame that doctors and members of the family weren't always with the dying," Worcester had insisted, "because of the reassurance to their faith in immortality that they would get from what they would see and hear from the dying." Cabot's own Unitarian religious sentiments led him along more materialist lines, and he expressed the idea that the revelations at the end of life might be more a result of failing brain cells. Yet, Ella at the end, he recalled, had seemed to die and then opened her eyes with a smile. "To see her face, such an expression—," he said. At the end of the interview, McCormick noted, Cabot finally wept.[52]

The Hidden Disease of Ella Lyman Cabot

Ella Lyman Cabot's final illness in the fall of 1934 brought to light deep ambivalence in her husband and in her physicians about the care of a dying person. She received medically informed, sophisticated attention to her comfort in the last weeks of her life, as well as the devoted attention of her medically trained husband. Yet she never discussed with her physicians the rationale or the goals of her treatment, apparently remaining ignorant about her lung cancer, the lethal prognosis, and the intentions of her caregivers—at least according to Cabot's account. Cabot reported that he had insisted that Ella be shielded from any mention of her cancer, and that she not know that her physicians had decided to forgo specific treatment for it. Her physicians undertook simply to ease her final days, "to just try to let her die quietly," as Cabot said.[53] Given Richard Cabot's public pronouncements about the need for absolute honesty on the part of doctors, it is not surprising that he took severe criticism from those close to him about this. His brothers Hugh and Philip both saw his behavior as evidence that his unpopular position on absolute honesty was practically untenable, even for him. Yet we might wonder just how effectively this strategy shielded Ella Cabot from an obvious reality. Ella's medical care served as a pretext, but perhaps not a deception—although Cabot's critics called it that.[54]

Ella died at Massachusetts General Hospital after nearly two weeks there in the care of the Dr. Roger Lee and with the cooperation and attendance of her husband. Lee knew that she was there to die. Although such an arrangement was unusual at the time, and was no doubt made possible only by Cabot's special status at the hospital, the motivation seems, in retrospect, commonsensical. Taking care of Ella at home had become impossible for Cabot. She would benefit from concerted nursing attention at Massachusetts General. Dr. Lee undertook to provide general supportive care for her there, while Cabot stayed in a room adjoining hers to be with her in her last days. Neither apparently discussed the nature of these arrangements with Ella.[55]

The events leading up to the hospitalization set the stage for the ruse. The couple spent the preceding August at their summer home in Cohasset, Massachusetts, where they had gone for a period of convalescence for Ella, who had been suffering from worsening breathing troubles that summer. During their stay in Cohasset, it became clear to Cabot from changes in his wife's physical state that she was growing worse, from a condition that he eventually diagnosed as lung cancer. As he recalled later, he woke one morning, "to find her sitting on the side of the bed almost crying saying she'd had a dreadful night and couldn't get her breath." He continued to treat her illness at their summer home and began to organize the days with the notion that they might be her last. Small routines emerged to help the couple through their time. "Right after lunch," Cabot recalled later, "she went up to her bedroom and lay down and I read her to sleep." Ella believed "that when anyone was sick the most important thing was to have a high spot in the day to look forward to and back on." Reading aloud gave the day its high point.

Cabot faced a difficult challenge in his dual role of husband and physician. As Ella's health began to decline during the summer of 1934, he examined her to try to sort out the cause of her problems. "I got a stethoscope and found her heart all right but her left lung all wrong," he recalled later. Putting this discovery together with other evidence, he had the answer: "I knew with almost certainty from all I'd done on differential diagnosis that it was cancer." After his discovery, he quickly determined not to let Ella know of it. It was not easy to hide what he knew, however. Soon Ella brought to his attention a surprising finding of her own, "something queer under my left arm." On inspecting the area, he found a large lump, "half as big as an apple," evidence of the cancer extending into the lymph nodes. Cabot determined to maintain his posture of unconcern. He couldn't remember that he "ever had to exercise greater self control in my life." He told his wife that it was a "big gland" and nothing more. In recounting this scene much later in a letter to his brother, Cabot sought some sympathy for his "situation, protecting her and at the same time pledging to tell no lies." He believed that what he had told her managed to satisfy the twin requirements and continued with the routine of daily care. The taxing effects of Ella's physical symptoms began to grow, however. Her appetite diminished, and she became nauseous. Cabot spent one whole night sitting up with her as she was vomiting and unable to sleep. He was, he remembered later, "pretty well played out" after that. With the help of a relative, he took Ella by train to Boston, where she was admitted to the hospital.[56]

The purpose and nature of the hospital stay quickly became evident, at least to the physicians. Drs. Lee and Cabot discussed the state of Ella's lung cancer. According to Cabot's memory of it, they "agreed on not trying to do anything at all like an operation." The purpose was to continue to provide the sort of nursing attention and symptomatic care that had become impossible for Cabot at home. The doctors at the hospital were in agreement to let her die quietly. Twelve days later, she did. Incredibly, Cabot insisted later that Ella understood from her hospitalization and the events leading up to it only that "she'd got very tired and was resting" in the hospital.[57]

Cabot's widely publicized and rigid position about medical truth-telling made his audience in later years unsympathetic about his predicament. His close relatives particularly saw Cabot's avoidance of the diagnosis of cancer in Ella's case as hypocritical. Hugh Cabot was notoriously impatient with his brother. Another brother, Philip, although not a physician, showed himself equally skeptical when he was given a hearing on the subject. In 1936, he brought up Cabot's withholding information about Ella's cancer in order to poke fun at his brother's "extraordinarily silly idea of telling the truth in medicine." Richard, he chided, had "deceived her as completely as any common liar like myself." But Cabot's brothers scorn was met with only denial, because Cabot resisted exploring the puzzling elements of his own story.[58]

Thinking back over the events of the long summer leading up to Ella's death, he reported that his wife "never suspected anything seriously wrong." He had succeeded in keeping the bad news from her, he insisted to McCormick in their interview two years after Ella's death. But did Ella really wake in the night breathless, discover for herself a large lump under her arm, and then decline steadily to the point of entering a hospital, where she received no evident treatment, and still not ever consider that she was dying? Who was fooling whom? Cabot reported that he never discussed the cancer with Ella. He had avoided discussing death with her and by implication avoided thinking about his painful coming separation from his wife and companion of forty years. Although Ella's hospitalization for medical care provided a pretext for her basic care, and Cabot's diagnosis provided him with a means of confronting a difficult fact about his wife's health, did this covert medical activity really keep at bay the recognition of death?

In her interview with him, Ada McCormick attempted to draw Cabot's attention to this hidden dynamic. She posed the obvious question. "You don't think," she broke in to ask Cabot, "that she knew and was sparing you as you were her?" Was it possible that Ella had conspired in allowing medical management to distract both Cabot and herself? They had cherished their last days together in part by avoiding a confrontation with the fact of her inevitable death. Cabot insisted that it was he alone who managed this charade. Could he admit years later that Ella had been his silent partner by hiding her knowledge from him? Perhaps Cabot saw that such an admission might rob Ella, retroactively, of the enduring power of her deception. If the fact of Ella's knowledge had been hidden this long, could it now come to light? The implications of Cabot's original denial were still too valuable for him to give it up. He had only one acceptable answer to McCormick's question. No, he replied, she never knew that I knew.

CHAPTER 7

FROM CABOT'S DAY TO OURS

Ideals of the Medical Relationship

This final chapter will offer observations about the path that we have traveled since Cabot's day in our ideals about how authority should be expressed in the relationship between physicians and patients. My goal in concluding is to bring into the present day the questions that I have raised in the preceding chapters and to link these questions to an ongoing debate about the nature of the modern patient-doctor relationship. This project faces an obvious challenge of scale in seeking to generalize broadly. So this final chapter will not sketch a history but examine a conventional account of recent events in light of the evidence from Cabot's practice and attempt to expand this conventional account by naming some useful added analytical terms. I shall also draw on new material that I think proves helpful in making the connection between Cabot's practice and our present circumstances. Cabot's records help document what his patients said to, and about, their doctors in the context of seeking care in the early decades of the twentieth century. The comparable experiences of patients in the middle of that century are still within living memory. I have drawn in part from a set of oral histories that I conducted over the past few years with older adults about their experiences of physicians' care in the mid twentieth century. These interviews offer a perspective with all the inherent limitations of hindsight. But such personal testimony helps to fill out our impressions of the remarkable transformation that occurred in the last third of the twentieth century. Together, these sources suggest that our conventional account of recent changes in the ideals for medical relationships may provide too limited a view.

Present-day writing on medical authority and medical decision-making suggests that Americans have revised their relationships with physicians considerably over the past thirty years. According to a conventional version of events, the usually dependent and submissive patients of the mid twentieth century have gradually adopted a more active stance in using their doctor's services, aided by advocates like medical ethicists and lawyers. Credit for these reforms rests with certain large-scale cultural and political changes in the 1970s that pressed patients to challenge a traditional model of the relationship that had long reigned. A willingness to confront accepted sources of authority and a new emphasis on individual rights inspired patients to take control of their medical care. Patients sought greater influence with their physicians. The efforts were diverse: the legal doctrine of informed consent created new powers for patients in research and clinical encounters, physicians grudgingly became more forthcoming about their procedures, while the American Hospital Association responded, promoting a "Patient's Bill of Rights" in 1972, for example.[1] One assumption underlying this account is that the hardships of illness and the mysteries of medical science naturally rendered patients passive. Control inevitably rested with physicians until mechanisms like informed consent restored a more equitable balance of power. Building on this assumption, the conventional story of the medical relationship has sounded a progressive theme. Reform in the patient-doctor relationship has been held up as an example, alongside civil rights, in which laws and regulatory policy have triumphed in the cause of personal liberty.[2] This is portrayed as a victory over the paternalistic power of physicians, promoting the ideal of patients as empowered agents in their use of medical care.

Informed Consent

In the setting of this conventional history of medical paternalism, Richard Cabot has seemed a harbinger of future change. In opposing a doctor's withholding a diagnosis from a patient, for example, Cabot was an early combatant in a struggle that was finally won in the late 1970s.[3] Cabot's concerns about the appropriate limits of individual medical authority seem to mirror contemporary concerns about patient empowerment. Although I recognize Cabot's famous dissent from medical authority, I draw different lessons from his activism. To start with, his patients seem to have been an active and engaged group, without need of any evident push in this direction from their doctor. Cabot seems to have been less a liberator than an ambivalent ally. Qualities in the contemporary ideals of the active patient were often to be found in Cabot's patients, like the woman, mentioned in Chapter 4, who wrote rejecting Cabot's suggestion that she see a particular surgeon for her bladder troubles because, she noted, "a Somerville lady was dilated by Dr. Morris Richardson [a surgical colleague] and could not hold her urine afterward." She had evidently been asking around to check about the recommended procedure. The personal burdens of illness and the complexity of technically oriented medicine did not render Cabot's patients uniformly passive. In the decades after Cabot, paternalistic models of the patient-doctor relationship did prescribe a very passive role for patients. But I would suggest that mechanisms a century later, such as informed consent, were an accommodation to change rather than its stimulus.[4]

The conventional account of patient empowerment nonetheless has an appealing ring to it and captures an important reality. Sweeping change in the 1970s affected the ideals that guided physicians and patients in reaching agreement about medical care. Considerable evidence for the change exists. The emergence of informed consent as a formal mechanism in medical practice provides an excellent case study. Putting oneself under the care of a physician in the 1940s or 1950s typically meant accepting or declining the physician's treatment without much further discussion. Acceptance of the relationship implied the acceptance of care. In the 1970s, however, a doctrine of informed consent took shape in clinical practice, drawing on changes in the regulation of medical research and driven by a series of civil court decisions on medical tort law. Informed consent specified that a patient would have to agree formally to procedures and treatments after receiving adequate information about the accompanying risks and benefits. Courts gradually expanded their interpretation of what constituted adequate information, giving patients greater control. No longer would a person undergo treatment by Dr. Jones simply because Dr. Jones was the doctor and recommended it. Now a patient accepted or declined medical care with a good understanding of its nature and after acknowledgment of its potential risks. This change was indeed a significant alteration in the basic processes of medical decision-making. Patients came to know more from doctors about their conditions and their options for addressing them. The conventional account accurately identifies this significant emendation to the ideals of medical relationships. At a minimum, informed consent placed procedural limitations on the doctor's ability to decide unilaterally about care on a patient's behalf.[5]

Still, this conventional account of the development of medical decision-making and informed consent in the late twentieth century leaves intriguing paradoxes unexplored. Certainly, there were changes in the formal and legal constraints on doctors and patients, and we have found an impetus for this change in broader cultural and political movements of the day that advanced personal liberties and autonomy. Yet one takes away from the literature on informed consent the odd impression of great advances in the face of much lost ground. Such reforms aimed at a form of control over medical care that may not have improved much for individual patients. Accounts of the creation of informed consent emphasize not only the doctor's unilateral authority but the daunting technical complexity of medicine and its installment in increasingly elaborate and impersonal settings like the modern hospital, the surgical suite, and the intensive care unit. Patients were perhaps inspired to challenge the medical system, securing mechanisms like informed consent to give them greater ability to influence their care.

But one might ask further how passive and disempowered patients suddenly acquired the ability to influence the behavior of their physicians. In the conventional account, it was allies like lawyers and bioethicists who helped them to win concessions. But how influential and how committed were such allies? If bioethics itself, for example, ever posed an oppositional challenge to medicine, it proved an accommodating opponent. Ethicists established a few secure posts outside of medicine in the later decades of the century, but they have always been as eager to take part as to take to task. Ethicists have tended, for example, to speak almost exclusively about the interests of patients already inside the medical domain, and not about the independent interests of the public in health. Throughout hospitals, medical schools, and research institutes in the late 1970s and 1980s, bioethicists secured a modest, but comfortable, place at the table. Did they really achieve this goal by browbeating their hosts into better manners? The conventional accounts of patients' rights represented ethicists as both more radical and more powerful than they seem in retrospect to have been. Ethicists played the role of concerned courtiers or cautious counselors to

American physicians attempting to negotiate the sweeping institutional changes of the late twentieth century. Bioethicists rarely found themselves outside beating on the castle gates.[6]

Certainly, informed consent became a law of the land in medicine. But I suspect that it did not alter the landscape so much as conform to other, less auspicious changes already afoot. While patients gained some leverage in the limited realm of their contacts with individual physicians, those contacts gradually became more attenuated in their ability to influence the overall process of care. Informed consent, I would argue, had as much to do with the latter as with the former.

A first improvement on the version of events sketched in the conventional account would be to note that much of the response to the pressures for patient choice came from within medicine. Mechanisms like informed consent have been presented as a successful external counterweight to the authority of physicians. But informed consent became a routine part of the protocols of medical practice, generated in part by physicians.[7] In addition, patients have rarely identified informed consent as an effective means to assert control over their care. People interviewed about the use of informed consent over the past few decades have tended to express the persisting belief that it originated with, and serves the purposes of, physicians.[8] In accordance with the doctrine of informed consent, physicians do indeed now enumerate the risks and benefits of their treatments in ways that they never did before, but the benefits of this change are far less than its proponents hoped. The American public has not come to characterize medical care as a more flexible or responsive enterprise in the decades since the 1980s. Informed consent has become an integral piece of a complex bureaucratic health-care system, not a solution to its hazards.

Certifying the Acceptance of Care

Informed consent promotes the ability of patients to choose, yet it does so specifically in the context of an increasingly segmented and specialized institutional system of American medical care. A spectacular case from the recent medical literature highlights some of the limitations of informed consent for patient empowerment as routinely identified by its many critics. The case also directs attention to the crucial institutional function that informed consent performs. As a means for expressing individual choice about medical care, informed consent can fail utterly, even when the performance seems to go ahead without interruptions or special concerns. Informed consent in such cases still serves as a convenient and efficient means for certifying the patient's acceptance of care, a process previously mediated through a patient's family and through the relationship established with an individual physician.

This case was reported in the *Annals of Internal Medicine* in 2002 as an example of how a series of small missteps in medical practice can result in massive error. The outcome in this case was shocking, but fortunately not tragic. A woman entered a large, academic hospital under the care of a team of highly specialized radiologists, who successfully treated her for an abnormal blood vessel inside her brain. The following day, while she was still in the hospital recovering, she found herself transferred accidentally to the hospital's cardiac electrophysiology laboratory. There she spent more than an hour under the care of a large team of highly specialized cardiac electrophysiologists, undergoing a risky and complex interventional test on the electrical conduction system in her heart. The procedure was intended for another patient, for whom she had been mistaken, having a similar name. A physician in the cardiac laboratory, prompted by a nurse, finally recognized the error, but only after electrodes had been placed in the muscles of the patient's heart measuring the effect of tiny electrical shocks that disrupted its natural rhythm. A critical event in this unfortunate story from our perspective was the use of informed consent. One of the physicians involved with the cardiac procedure met with the woman before the procedure to get her informed consent. He brought the necessary forms, which this woman dutifully acknowledged and signed, seeming to agree that she understood and consented to a procedure that no one meant for her to have. The woman faced no obvious barriers in language, socioeconomic background, education, or disability that should have hindered communication with her several physicians. Ironically, her own daughter was a physician.[9]

The use of the informed consent documents could, of course, have been the occasion for a detailed conversation about the nature and rationale of the heart procedure, with the likely consequence that the mistake would have become clear. But neither the physician nor the patient seems to have recognized at the time that informed consent had failed them. Both parties were familiar with the basic routines of consent in the hospital. The patient had almost certainly given informed consent for the radiological procedures the previous day. Neither she nor her physicians found any reason to question the proceedings, and they seemed adequate for their purposes. Informed consent had, in fact, accomplished an important step in formally certifying the patient's acceptance of her new physicians' care, although for the wrong procedure.

Informed consent hinted at no other obligations of the physicians to do more in arranging, following, or promoting this patient's medical care. It served at best as a weak contract for service. The relationship to any identifiable physician in these arrangements was minimal, with obligations and interests defined by the nature of the technical service being provided. Risks and benefits were proposed and acknowledged in some manner. The patient signed on the dotted line, showing acceptance. Remedy for the errors that were exposed in this case might have included tighter surveillance of the use of informed consent. Clearly, the patient cannot be counted to have received adequate information about the procedure. It was the wrong procedure, after all! Knowledge of the details of what was being offered would have given her better power to decline—important in this instance. But even impeccable use of informed consent provided only this minimal ability to avoid a particular service being proposed.

The editors who selected this case for their journal invited a panel of experts to comment on the errors that it revealed. One of the commentators confessed to an analogous concern about the limited effectiveness of informed consent. His initial reaction on considering this case, he observed, was "to advise patients to become as well informed as possible—to know exactly which medications they are taking and which procedures are planned—in order to protect themselves." He recognized on deeper reflection, however, that more might be required: "But my other reaction is that's absurd. Why should we have to rely on patients to protect themselves?"[10] The discussant moved on to consider how the hospital might better organize its facilities to avoid such errors. The personal relationship and obligations of the physicians to this patient hardly figured into the analysis at all. Physicians were simply one specialized element in a larger system that was under scrutiny.

Informed consent partially replaced older models for certifying the acceptance of a physician's care. A useful thought experiment might place a similar patient in a similar situation a half-century earlier. Confronted with an unknown doctor who was offering some new medical procedure in the hospital in 1950, the patient would be unlikely to discuss the accompanying risks or benefits, or even entertain any lengthy discussion of the procedure at all. In an older paternalistic model for medical care, she might have expected that detailed choices about medical care were not required of her in this setting. The response of a patient to an unknown doctor offering a new procedure in the hospital of the 1940s or 1950s might have been: "I don't know, ask my doctor," or perhaps, "You should talk to my family about that." The expectation on the other side was that any new physician would be required to channel care through a physician with an established relationship. Established relationships in turn included the family among those who shared responsibility for medical care. Although the process of certifying approval for medical care in the era of paternalism did not necessarily involve the patient, it could be complex and even protracted.

The Portability and Efficiency of Informed Consent

Informed consent, in giving greater autonomy to patients, also proved to be a successful adaptation to the segmentation of care in complex medical institutions. Carl Schneider, a legal scholar concerned with the pragmatic limitations of informed consent and the autonomy model for patient-doctor relationships, has noted parallels between the effects of bureaucratic segmentation on medical care and the effects of autonomous patient decision-making. Both the ideal of autonomy and the model of bureaucratic efficiency, Schneider contends, contribute to medical relationships that are less personal, less dependent on trust, and less mutually involving. Autonomous patients limit the control of their physicians, but they may also miss out on more sympathetic involvement.[11] Schneider emphasizes how both patient autonomy and medical bureaucracy will tend to increase the impersonality of medical care. But he does not draw what I see as a more obvious connection. The mechanism of informed consent, even if it does not reform the medical relationship appreciably, still serves an important bureaucratic function.

Informed consent has provided a wonderfully portable and efficient means for approving the acceptance of medical care in a highly segmented and specialized medical system. The power to give informed consent today moves along with patients through medical institutions. It is portable. Specialized physicians who meet patients along their paths through the sites of care can make agreements about procedures quickly and serially. A specialist approaches the patient in the radiology suite, the emergency department, or recovery room. He describes the risks and benefits of his service and secures consent. Older models for the certification of medical care were far more cumbersome in such circumstances. Approval for care in mid-century medicine, although unburdened by formal procedures, may actually have been more cautious. Care had ideally to be coordinated through existing medical relationships with a single physician. Such arrangements posed a challenge to elaborate networks of specialized practitioners working in segmented institutional niches. Paternalism invested the physician with great control and independence in deciding about care. Yet the authority to manage this responsibility was firmly vested in the person of the physician who established and maintained a relationship with the patient and family. This authority might be conferred briefly on physicians outside of established relationships, especially in an emergency, for example, but it would not typically be extended far without creating the presumption of an ongoing responsibility for care. In modern bureaucratic institutions, such personal, charismatic relationships would be increasingly difficult to maintain, as the sociologist Max Weber had already noted at the beginning of the century.

Informed consent narrowed decision-making powers efficiently to the individual patient. Any physician could be involved in a given instance of consent. In addition, the patient's family could often be efficiently dropped from the group needed to approve care. The older paternalistic model, in contrast, required coordination with the patient's family. Cabot's colleagues emphasized this point when they insisted that agreements should be made with the family alone about the care of patients with life-threatening disease. The gradual diminution of involvement of families in late twentieth-century medical care was often overlooked, or even suppressed, in the early critiques of paternalism by advocates of greater patient autonomy. This early critical literature of the 1970s tended to emphasize the goal of redressing imbalances of power between physician and patient. Attending to the accompanying shift of power away from families and other caregivers would only have served to distract from this proposed goal.

Debate over the disclosure of a diagnosis of cancer to a patient provides a serviceable example of the willful suppression of the complexity of paternalistic medical relationships. Survey data in the United States from physicians in the 1950s and 1960s show that they held views on the diagnosis of fatal disease that were roughly similar to those of Cabot's peers. Physicians generally agreed that information about the diagnosis of cancer should not be disclosed too readily to a patient, *pace* Cabot. But they were equally adamant that it was inappropriate to hold on to such information unilaterally. When American physicians excluded a patient from information about a life-threatening disease, they tended to do so in favor of consulting the family. Cabot had documented, and opposed, this practice among his peers.[12] Yet the critique of paternalism that arose in the 1970s tended to represent prior medical handling of such information as strictly unilateral, even to the degree of ignoring as inconsequential the physicians' discussion about the involvement of families.

Celebrating the overthrow of medical paternalism, physicians in the late 1970s reviewed the data on their erstwhile crimes. Between the 1960s and the 1980s, surveys indeed indicated that American physicians had completely reversed their opinions on the disclosure of a diagnosis of cancer. A large survey conducted in late 1959 and 1960, for example, determined that roughly 90 percent of physicians felt that they should not tell a patient about a cancer. This finding supported the findings of a similar earlier survey.[13] In 1977, a group of physicians, sensing that change was in the air, conducted a similar survey that compared itself to the prior results. It demonstrated a marked reversal: 77 percent of physicians now stated that they would disclose a diagnosis of cancer to the patient.[14]

In reviewing these data in 1979, the researchers observed with evident pleasure that new respect was being shown for the patient's interest in guiding care. The review emphasized the shift in power that had occurred when physicians began to give patients the information that they had previously withheld for themselves. This critique neglected to note that physicians in the original survey had routinely insisted that the information was not kept as a medical secret. In both of the large, early surveys from the 1950s and 1960s, physicians had repeatedly insisted that, of course, "a responsible member of the family should always be told."[15] The ideal was not to maintain exclusive control of the information but rather to share it with people who might be called upon to approve care, the family members. By stressing how physicians had withheld information from patients, the reformers of the 1970s emphasized that individual rights had been restored. They neglected to ask whether information was being shared with the family at all. The successful defeat of paternalism cloaked a more complex shift that had occurred in the processes of certifying the acceptance of medical care.[16]

Patient Power and Consumer Choice

Patients and physicians in the final decades of the twentieth century met increasingly in segmented, specialized, and compartmentalized sites for care. During a single episode of illness, a person might move quickly from office or clinic to emergency department, radiology suite, hospital ward, intensive care unit, cardiac catheterization laboratory, and operating room. Such arrangements were not new. The complexity of medical institutions has grown steadily since Cabot's day. But the final decades of the century brought a unique shift in the response of American physicians. During the 1970s and 1980s, physicians began to acquiesce to something that they had previously avoided and increasingly to define and limit their practices according to the institutional site of service. Emergency medicine became a distinct specialty during the 1970s, while intensive care units over the same time became a limited area for specialist practice.[17] Rehabilitation hospitals and nursing homes gradually developed a cadre of specialized physicians, dedicated to exclusive institutional practice. And in the last decade of the century, the specialty of internal medicine notably split off a group of physicians specializing in the management of general hospital wards, calling themselves hospitalists and limiting their practice to the care of patients only during the time of hospitalization.[18] Encounters between physicians and patients have consequently become gradually more episodic and more strictly defined by the institutional site.

Informed consent met the needs of more finely partitioned institutional structures. As patients passed under the attention of a succession of specialized physicians in different sites, they were greeted with a fantastic array of medical services: an injection of medicine, a dye-enhanced x-ray, placement of a plastic catheter in the vein of the neck, or surgical excision of the gallbladder. Modern emergency rooms, cardiac care units, surgical suites, hospital wards, and recovery rooms left little space to organize the traditional groups required for certifying decisions about these procedures—whether under the prompting of physicians or of patients. Patients received more procedures by more different specialized practitioners, and rapid clinical decisions were necessary in order to deliver the increased volume of services. The traditional constituency for medical decision-making of the primary physician, the family, and the patient was difficult to bring together repeatedly to review the progress of care and to authorize new steps. Patients had to become more efficient consumers of health services, and informed consent was a model for decision-making that enhanced this ability.

Informed consent and related mechanisms may have permitted patients to exercise more autonomy and control, but patients did not thereby acquire powers that were new or revolutionary. Instead, they enhanced a well-established power that they had long held as consumers in the medical marketplace.[19] Acting on informed choices about services is what consumers ideally do. Modern patients have had a version of this capacity for as long as medicine has been part of the modern economy, although they have been able to express it better or less well as the marketplace and medicine have changed. The discussion about empowering patients in medicine moved in tandem with the deployment of the notion of consumer rights in medicine, at least in the early phases. Consumerism received new attention even as the doctrine of informed consent was providing a mechanism to put it into effect. The medical subject heading of "Consumer Satisfaction" appeared for the first time in the American medical literature in the 1960s as a standard means of categorizing articles pertaining to the relationship between physicians and patients. The comprehensive *Index Medicus* introduced this formal subject category coincident with the first round of late-century debates over the empowerment of patients in decision-making. An early statement on the need for greater patient autonomy by the sociologist Eliot Friedson appeared in 1960, for example, indexed in the medical literature under the new heading "Consumer Satisfaction." The power as a consumer was not as profound as control over the content of health practices or access to the proper treatment. It was a circumscribed ability to examine and to choose what was made available.[20]

IN LIGHT OF THIS ANALYSIS—and through an accident of history, perhaps—Cabot's patients might seem oddly familiar again. Physicians have at times wielded a rather unalloyed authority over their patients, but that era seems to have followed Cabot's and to have preceded our own. Perhaps we have come full circle in the expression of the patient's capacity as a discriminating client of the physician's services, from a day when American physicians could not yet presume to have the unmediated allegiance of their patients to a day when they could no longer do so.

In the early twentieth-century medical office, Cabot's patients exercised similar kinds of choice about medical services, although in a different context. The patients who frequented his office were often quite active and opinionated about their medical care. They selected Cabot from a variety of practitioners, whom they sometimes reviewed for him in detail; they paid directly for these services; and they had some limited ability to negotiate the appropriate fee. Although these patients labored under significant burdens of illness at times, they demonstrated an eagerness to make choices about the handling of their problems. They sought information and advice to guide and support their choices. Their activism sometimes even trumped a natural deference that they may have felt toward a well-known, elite physician like Cabot. A former farmer from the western part of Massachusetts had no reticence, as we have seen, in writing to tell his eminent doctor that "you made a bad guess." Even a patient who promised that she would be "faithful in following all that you recommend" arrived accompanied by a letter from a prior physician indicating that she had unceremoniously rejected all previous medical advice and treatment. Indeed, she soon wrote back to Cabot after a first visit challenging his diagnosis and wondering whether he had something better to offer. People wrote to express the sense that they had to be scrupulous in weighing their choice about care and even seemed on occasion to be "shopping" in their quest for assistance. They expressed an interest in choosing carefully what they accepted, and sometimes moved on when the promise of help seemed stronger elsewhere. Many people did go along with the recommendations that Cabot made; but when Cabot managed to carry his point, he often seemed to win out through persuasion and persistence rather than authoritative decree.

If we have come full circle in our expectations about patient-hood, first to greater passivity in the middle of the century, and then back to activism in the late century, then these changes must depend on more than the debilitating consequences of illness or the technical mysteries of medicine. Patients today are no less burdened by the suffering, disabilities, and indignities of ill health; in fact, people may live under a greater burden of multiple, interlocking, chronic illnesses today than they did earlier in the century. Elaborate medical technologies, while mystifying at times, can prompt more caution and skepticism than they do awe and acquiescence. Patients have not recently become more expert in seeing through the mysteries of medical science and technique. These basic asymmetries in relationships between physicians and sick people seem to me rather similar from Cabot's day to our own. Indeed, the rising activism of patients in the past few decades noted in the conventional account took place in the setting of ever more fragmented care and more esoteric technical practices. Yet people often seek explanations, reassurances, and justifications as much as they seek treatment and relief. A broader cultural challenge to authority in the late twentieth century may have inspired patients and their advocates. But the conventional account of recent change in medical relationships has inclined us to view relationships in terms of a balance of power. This analytic frame has been adequate for the purposes of advancing reform, but it has skewed our ability to appreciate institutional sources of change in late-century relationships.

Doctoring among Patients

In the mid twentieth century, physicians promoted a paternalistic model of medical decision-making that gave them considerable individual influence over the care of their patients. Cabot's colleagues perhaps shared this aspiration a couple decades earlier, but it required the growth of the physician's general social authority to allow it to function. We know a considerable amount about the institutional and professional advances that supported this growth in authority. But the discussion here has emphasized the parallel role of patients in ceding, rather gradually and perhaps temporarily, their capacity to act as discriminating consumers of medical services. Were there features of the early-century delivery of medical care that prompted patients to seek shelter from the decisions of consumers? The technical mysteries of disease-oriented care certainly contributed. But for some of Cabot's patients, it proved daunting to shop for health services. They even had a colloquialism to express this problem: doctoring.

The early twentieth-century market for medical services challenged Cabot's patients on occasion in a way that is vividly portrayed in their letters. Being a consumer of medical care in the early century posed significant difficulties. This plight was sometimes quite affecting. In 1916, a young woman wrote describing a taxing series of encounters with doctors. She suffered from a disabling pain and swelling in her limbs and joints, she explained in her letter. The condition had progressed so terribly that she had become unable to dress or bathe herself. She recounted a list of diagnoses and treatments that she had already obtained from different doctors, all in vain. "One doctor said neuritis," she explained. "My family physician said it was arthritis. Then Dr. Burgess said it might be a tooth abscess. He x-rayed my teeth and took care of an abscess. A throat specialist then extracted my tonsils." Each specialist had seemed able to deliver a service, but none brought much improvement or relief. Still hopeful, she had widened her search: "I wrote to the Mayo brothers, but they said not to come as they would do nothing for me."[21] She now sought Cabot's advice about her most recent encounter. She had found a new doctor who, she reported, "says it is not arthritis but hookworm" and offered to diagnose and treat her case for a fee of $1 for the x-ray and $4 for the medication. She explained the quandary that she found herself in. "I have no way of knowing whether the doctor is right. He told me more than any other doctor did, and used the x-ray, through which he said he could see the worms." It sounded promising, but she felt uncertain about what she would actually get for her $5. She concluded her letter asking how it was possible "in this day of such wonderful things as are being accomplished in medical science that there is no help for me?"[22] What constituted a proper search for medical care, and what helped to distinguish among the many options available to a person suffering from ill health? New and wonderful things appeared continually on the horizon. Was it simply a matter of *caveat emptor,* as this correspondent seemed to fear? She appealed to Cabot for guidance.

This woman found herself facing a dilemma, debilitated by illness and confronted with a daunting number of avenues for treatment. Her account hinted too at an ironic edge in the predicament: that more treatments seemed only to leave her worse off. In one sense, these kinds of problems had been evident to experienced patients and physicians for a long time. The hazard of too many possibilities for treatment and too few opportunities for real cure was often commented upon. In the seventeenth century, Robert Burton's *The Anatomy of Melancholy* warned those who frequented many different doctors and "try a thousand remedies, and by this meanes they increase their malady, make it most dangerous."[23] The woman mentioned above wrote to Cabot to describe a twentieth-century version of this hazard. The removal of tonsils, the excision of a dental abscess and the irradiation and chemical eradication of hookworms entailed a level of technical complexity and bodily invasiveness that was increasingly characteristic of twentieth-century medicine. The growth of medical resources had amplified a long-standing problem.

If versions of the problem had long been evident, however, its twentieth-century variant seemed to carry new urgency. Evidence scattered through the correspondence of Cabot's patients and their peers in early twentieth-century New England suggests that the choice among a plethora of different medical services posed a novel challenge. A new colloquial usage emerged that captured the anxiety of patients who sought hard for medical care in an increasingly complicated marketplace. Both Cabot and his patients occasionally used the term "doctoring" in their informal notes and letters in a sense that was particular to the times. Doctoring meant repeatedly seeking out and trying medical services, with the implication that they were not helpful. In the common language of the day, "to doctor" meant, as one lexicographer noted precisely in the pages of *American Speech* in 1927, "to have been treated by several doctors with the implication of no improvement."[24] It was a new word for a familiar irony. This particular colloquial use of "doctoring" shows up distinctly in the letters of patients writing to Cabot.

Writing in 1902, a middle-aged woman who had been through a course of unhelpful treatments for her illness confided, "I am better now than I was April 7 when I stopped doctoring."[25] Doctoring, in the sense of visiting many doctors and receiving many treatments, had not helped her. A different patient, also writing in 1902, provided a more detailed account of a long series of futile medical encounters. She suffered from headaches, swelling in the belly, and episodes of "very sick times." She wrote having recently arrived back from a visit to a physician in Buffalo where, she reported, she "was treated by electrolysis." She mentioned the practices of four different doctors by name, as well as an anonymous "herb Dr." whom she had consulted. Her return from Buffalo, she decided, "was the last of my doctoring." Unfortunately, she noted, "I cannot see that it helped me very much."[26]

Doctoring, in the accounts of patients, was a trying and unproductive enterprise. In manuscript collections similar to Cabot's records, the same use of "doctoring" as a patient's action appears in correspondence. Physicians at the Massachusetts General Hospital occasionally tucked letters from their patients into the medical records of the hospital. One such letter, from a woman, speaks of belly pains throughout the previous year, although she had "doctored off and on until the next July."[27] Doctoring had brought her many services but little relief.

The verb "doctor" used in this sense appears in the written record of the period only in manuscript collections of informal notes, dictionaries of colloquial speech, and dialect fiction that aimed to capture the common speech of the day. Edith Wharton's novel *Ethan Frome* (1911), for example, which will be familiar to many readers, picks up exactly this usage in the speech of local characters in turn-of-the-century New England, who describe the behavior of one of the novel's main characters, Zenobia Frome, as "doctoring." Throughout the novel, Zenobia struggles with a vague, debilitating illness, and, as one neighbor puts it, she has "always been the greatest hand at doctoring." She is constantly going off "to Bettsbridge or even Springfield to seek the advice of some new doctor," from which trips she "always came back laden with expensive remedies and her last visit to Springfield had been commemorated by her paying twenty dollars for a battery."[28] The novel turns on events following Zenobia's departure for the office of yet another doctor, this time suggestively named Dr. Buck. During her absence, her husband is injured in an accident, which kills his mistress, and winds up an invalid, dependent ironically on Zenobia, who discards her doctoring for a newfound capacity as his caregiver.

The colloquial roots of this usage may be glimpsed in earlier literary appearances that shed some light on the particular nuances of the word. *Webster's Imperial Dictionary* of 1904 gives "to subject oneself to medical treatment" as one meaning of the verb "doctor" but leaves unstated the identity of the person providing the treatment.[29] This sense of doctoring seems to have derived from an earlier nineteenth-century colloquial use of the term to mean treating oneself in the absence of a doctor. This picks up on the notion of personal responsibility for treatment, which had deep roots in the American traditions of self-medication and domestic care.[30] A short story printed in *Harper's Magazine* in 1880 by Sarah Orne Jewett, whose father was a physician in rural New England, illustrates this earlier colloquial use of doctoring as the action of a patient when there was no doctor. The narrator of the story frames it as a re-telling of a tale that she had heard during an accidental encounter on a country outing. The story tells of a farm woman, "young Joe Adam's wife," who is "dreadful nervous," and "forever a-doctoring," taking draughts from the contents of "an old green boardbox that was stowed full o' bottles that she moved with the rest of her things when she was married." She sees no doctors in the story, merely telling a local minister "how many complaints she had, and what a sight o' medicine she took."[31]

In the early twentieth century, with a rapidly growing range of medical practitioners available, one colloquial sense of "doctoring" shifted. Doctoring reflected anxieties over the search for good therapy when the responsibility rested with the sufferer. Nineteenth-century self-medication grew into the task of choosing a doctor's treatments in the early twentieth century, and it was understood that a longer search sometimes yielded a smaller final benefit. The role of the consumer in the medical marketplace was on occasion a difficult one. Colloquial use of the term "doctoring" helped express concerns that faced a sick person confronting a practitioner who, for example, offered to treat hookworms with an x-ray and a bottle of medicine, for a prearranged fee of $5.

Only a few of Cabot's patients expressed worries about doctoring, perhaps not surprisingly. In part, the behavior of doctoring implied a very select kind of illness, one that was persistent, troubling enough to drive a search for care, but not so debilitating as to keep the sick person from taking responsibility for the search. In part, too, typical arrangements in Cabot's office accommodated a more congenial sort of consumerism and helped to transform consumerism into another kind of relationship. A physician who undertook to guide, monitor, and mediate the search for specialized medical care might lessen the anxieties of doctoring. Chapter 2 has noted that Cabot combined the practices of consultation and referral as different means for connecting his patients to other physicians. The newer arrangements of referral especially seemed a reasonable corrective for the problem of doctoring, as a single doctor became a means to filter and certify the use of multiple, specialized practitioners.

A system of referrals and an insistence by physicians on maintaining the medical relationship in the face of multiple different providers and multiple different specialized services were characteristic of the medicine that emerged in America in the decades after Cabot. The emphasis on a single, paternalistic relationship blunted the edge of consumerism in the exchange of medical services. Patients moved by referrals among different practitioners through a system that was structured to preserve the significance of one primary relationship. Paternalism substituted for consumerism in a model of doctor-patient interactions that specified stable mutual responsibilities and duties within a hierarchical relationship. We have inherited a caricature of these relationships in the critiques of paternalism that developed during the 1970s. But more detailed evidence of their particular qualities can be found in other sources.

Goods and Gifts

While the patients in Cabot's office appeared to act on occasion as consumers, the changes in medicine that they experienced tended to hinder the expression of consumer choice. The complexity, invasiveness, and risk of twentieth-century medical care constrained Cabot's patients and freighted their therapeutic choices with special burdens. The technical content of medicine never seemed entirely opaque to patients, but the burgeoning range of special services and techniques in the middle twentieth century made choosing among them increasingly intimidating. A patient's acceptance of a physician's care in 1930 or 1950 might well have implied the acceptance of treatment and intervention. Refusal of a procedure, test, or a medication was tantamount to rejection of the physician and might necessitate finding care elsewhere. Choosing a physician was not a process to enter into lightly.

Before considering the testimony of people about their mid-century encounters with medical doctors, we might consider the alternatives to a consumer relationship. The relation of the consumer to the marketplace is so familiar to us that it seems difficult to conceive of its counterparts. Yet patients and physicians alike often resist the model of consumerism for the exchange of medical goods and services. Calling the counterpart to medical consumerism medical paternalism really leaves us no farther ahead. Calling it merely acquiescence to professional authority shifts the focus away from the agency of the patient and individual experiences of medical care. Existing theoretical structures in anthropology offer another model for exchange relationships that are clearly distinguishable from the model of consumerism. Anthropologists in the early twentieth century conceived of two basic types of economic relationships, which played out in different social contexts, calling them gift exchange and commodity exchange. This distinction provides another axis for analysis of the exchanges between physicians and patients, separate from the balance of power and control. Because anthropologists used the word "gift" originally in a narrow sense that can be confusing on first encounter, it is convenient to refer instead to the difference between traditional markets and modern commodity markets. The French anthropologist Marcel Mauss originally noted that economic practices in different cultures differed strikingly from his familiar European models. What I shall call traditional markets seemed to Mauss to function in a way that was at odds with modern economic markets in the industrialized sectors of the world. In comparing the nature of transactions in traditional markets, anthropologists, starting with Mauss, formulated a general scheme distinguishing what they called gift exchange from commodity exchange.[32]

This difference can be expressed in a simple vignette. If an important leader in a traditional social system gave you a pot of rice, that rice carried a greater value than an identical pot of rice given by someone of lower status. Upon receiving the rice in the context of a traditional market, you would not be expected to give back something of identical value; in fact, such a response might be perceived as an insult, demeaning the value of the gift and the relationship offered. Instead, you would attempt to repay the giver sometime later, perhaps at an unexpected time, with a countergift of different and perhaps greater value, say, a wool blanket. Special obligations were entailed in these exchanges, in part because the rice came from someone of higher status in the village. The relationship between you and the important leader, as recipient and donor, would be sustained in an ongoing exchange of such gifts, which were valuable goods in the traditional market. There was nothing charitable or frivolous about these exchanges; they were the lifeblood of the local economy and provided one with a means of distributing goods among the members of the society. The routine rituals of giving and receiving defined this traditional market.

Several important features of traditional market exchanges became apparent to the anthropologists who studied them in the early century. The materials and services that were traded in such markets seemed to carry a value that derived in part from the identity of the people who gave and received them. In a modern market, goods have a value that is abstract and independent of the identity of the buyer and seller. The things being

exchanged in traditional markets, however, were more like personal possessions than abstract goods, and the anthropologists called them gifts—
although they were quite different from gifts in their usual modern sense.
These traditional gifts carried with them the identity and memory of the
giver. The rice that you received from the village head was more valuable in
coming from a powerful source. The exchange of such personal possessions in the traditional market also occurred in a context of an ongoing,
often hierarchical relationship. Traditional gift transactions entailed continued, often asymmetrical, obligations that were foreign to modern markets. The assumption behind exchange in the modern market is that
participants will settle their accounts fairly, with exact and timely payment.
In the traditional market, the transfer of goods was never complete at the
time of the exchange but instead created new obligations, which were reaffirmed and strengthened by subsequent exchanges. To "settle the account"
exactly and quickly in the traditional market would mean to reject the relationship and so insult the donor. Traditional markets functioned through
the establishment of sustained, mutually obligatory relationships. In the
modern market, in contrast, the relationship between the buyer and the
seller exists simply for the purposes of the exchange and ideally does not
affect the value of the goods.[33]

To make an analogy between gift-exchange relationships and medical
relationships requires some additional observations taken from the anthropological literature. Money, it must be noted, can also function in traditional markets; and barter can likewise take place in modern markets. So
paying the physician with a chicken or a bag of potatoes does not in itself
constitute a gift-exchange relationship. Gift exchange is about the nature of
the relationship between the buyer and seller rather than the currency in
which they trade. Money is one such currency. While monetary valuation is
the abstraction that makes modern markets efficient, money itself can be
used in other ways too. Cash can, for example, function in the manner of a
traditional gift if some of its value derives from the person of the donor—
as, for example, when the value of a monetary reward includes the honor
associated with the prestigious person who bestows it. In addition, even in
modern markets, money is an indicator not just of wealth, but of power
and prestige, which are qualities that attach to the person who holds
money. Even in traditional markets, a gift can be assigned a monetary
value, although the attempt to settle an account by returning the exact

monetary value of a gift would, of course, be to ignore the gift's most important qualities.

Barter may, in a comparable way, function equally well in modern commodity exchanges. Paying for the doctor's services with potatoes could be a form of market exchange, as long as payment implied no ongoing hierarchical duties or obligations in the relationship. Modern markets require that the value of goods be set abstractly through the balance of interests between the buyer and seller. But goods can be bartered as commodities in a modern market without the creation of a traditional gift-exchange relationship. If the trade of the previously mentioned wool blanket and rice assumed and entailed no continuing obligation between the donor and recipient, and depended solely on the assigned value of these items, then it would be in effect a modern commodity exchange. One distinguishing feature of the goods exchanged in commodity markets is that they are fungible. An object that is fungible can be replaced, in part or wholly, by another object of equivalent value. Gifts can never be treated as fungible, in part because the value of a gift depends on the nature of the relationship between the people who exchange it. In contrast, the people who make exchanges in a modern market are in a sense fungible with respect to one another. The buyer or seller can be replaced by any other individual who is willing to make the same exchange at the same price, without affecting the transaction. Commodity-exchange relationships exist transiently with no preceding or continuing obligations and no inherent imbalance of power, whether the exchange involves money or barter. The goods and services exchanged have no personal, identifying connection to the parties exchanging them: that is to say, they are alienable from the relationship.

The interactions between doctors and patients in the twentieth century sometimes shared the qualities of traditional gift-exchange relationships, even though these interactions took shape in a modern market economy. Crucial distinctions must be preserved in order to make these analogies, however. One pitfall of applying the model of gift exchange in a modern context is the confusion of different forms of gifts in such a context. In the setting of a modern economy, traditional gift-exchange relationships can be adopted to alternate uses. Even the term "gift" has assumed a significance that is at odds with the specific meaning originally assigned to it by Mauss. The exchange of personal tokens between patient and physician does not in itself designate a gift relationship in this classic sense.

In modern economies, the exchange of gifts often carries a novel symbolic value as a bulwark against the depersonalizing forces of the commodity-exchange market. But these symbolic gifts differ in significance from gifts in traditional market exchanges. To maintain the precision of this terminology, the anthropologist James Carrier has adopted the term "present" to indicate a modern symbolic version of the traditional gift.[34] So when a patient gives a present to a physician, for example, the exchange has only passing relevance to the nature of gift-exchange relationships. Presents in the context of modern markets are idealized forms of traditional gifts, which offer a partial remedy to the problem of personal relationships that become attenuated or overwhelmed in an impersonal market. A present in this sense is a kind of value-free, nonobligatory gift, offered as an antidote to the depersonalization of modern economies. Such presents differ from gifts in the classic sense, in that they are given outside any market and without market obligations of any sort. People who give and receive presents in modern economies insist that their true market value remain hidden and unquestioned. "It is the thought that counts," we say of the ideal present. Like its classic counterpart, the gift, the modern present continues to carry with it the memory and identity of the giver, but with an almost opposite effect. A present reverses the intent of a traditional gift. Rather than imposing obligations on the recipient, a present purports to represent the obligations of the donor, to say, in effect, "This object shows my concern and affection for you." Like that of a gift, the value of a present is partly dependent on the nature of the relationship between the donor and the recipient. But whereas the traditional gift formed the unit of exchange in a traditional market, the modern present putatively exists outside the market, as a response to its flaws.

Much about medical care in the twentieth century encouraged the development of relationships that were personal, nonfungible, and involved ongoing mutual obligations. In other words, relationships between physicians and patients readily took on some qualities of traditional gift-exchange relationships. The exchanges between doctors and patients trafficked in the most profound insecurities of the human condition. Greater control over the processes of disease meant that physicians involved themselves more deeply in the vulnerability, health and dysfunction, life and death of their patients. Although all expertise involves the handling of risk in the modern world, medical expertise began to claim primary responsibility with respect to risks that have a profound impact on the person of the client, the patient. Much about the daily work of medicine reinforced these qualities. It was hands-on, direct, and physically intimate for both doctor and patient, in separate ways. Exacting diagnosticians like Cabot inquired in detail into the most private aspects of personal life. Commonly used tools like the cystoscope and vaginal speculum displayed the intimate, interior spaces of the body, necessitating close contact between physician and patient, albeit in a formalized, technically mediated fashion. Such activities involved high levels of interpersonal trust, much as they required trust in technical expertise. Patients often gave bodily substances as part of their exchanges with the doctor. Substances like blood or urine might seem an extension of the person who gave them. Depending on the context, allowing a physician to draw a sample of blood seemed a powerful gesture of trust and connection. Such products were inalienable, we might say, or more difficult to disentangle from their personal identification with the giver in a commodity market. Even though exchange relationships between doctors and patients developed in the setting of a market economy in America, physicians struggled to limit the potential fungibility of their services, and to nurture a sense of mutual obligation with patients that extended beyond the mere discharge of debt.[35]

The anthropological analysis of gift exchange encourages us to see the shift in power in the medical relationship in a different light. Exchanges in traditional markets accommodated asymmetrical, hierarchical roles for the donors and recipients, like the relationship between the paternalist physician and patient. But modern markets rely upon more egalitarian relationships. While a gift from the leader of the village always carried greater worth, abstract valuation of goods and services in the modern market separates the value of goods and services more comfortably from hierarchical imbalances between the parties to the exchange. The attempt to recast the ideal medical relationship as nonhierarchical, negotiated, and even contractual marked a shift, then, along an axis from the gift-exchange model to the commodity-exchange model. In part, such a change stemmed from a late twentieth-century desire to remedy imbalances of power, as we have noted. But simultaneous change in the institutional and bureaucratic context of medical practice has supported a conception of the patient as a consumer of medical services, and this model of the relationship in turn promoted the recent alterations in ideals of the relationship.

Gift Relationships and Family Obligations

Observations about the nature of relationships with physicians in the mid twentieth century are abundant. The analysis that follows derives from a selective, but fascinating, series of oral histories conducted to document the recollections of older people about long-standing relationships that they maintained with physicians in the mid twentieth century. The discussion here serves as an example of how such materials might inform our understanding of the transformation of patient-doctor relationships over the later twentieth century. These open-ended interviews resulted in roughly twenty hours of transcribed accounts from nineteen older men and women living in and near Kansas City, Missouri, in 2000. The average age of those interviewed was seventy-five, and the average duration of their relationship to one physician was nineteen years.

In these interviews, people described their earlier experiences with medical care, focusing on physicians who had provided them with care for a large part of their adult lives. The interviews contained fascinating observations about the experience of being a patient in relationships that might be characterized today as fundamentally paternalistic. Most significant to this analysis were the general observations that people offered on the ways in which these relationships started and how they were transferred from one physician to another. People described their connections with their physicians as having qualities reflected in the model of the gift-exchange relationship outlined above. They tended not to talk about their use of the physician's services as though they were consumers of medical care. Their relationships were typically obligatory, certainly hierarchical, and carried a value that was dependent on the nature of the personal connections involved.

The most striking feature of these relationships was the matter of family. Interviewees often described their relationship to their doctors in terms of kinship. The doctors were noted to be "almost like family." These relationships thus seemed obligatory, with duties like those owed by family members. One eighty-nine-year-old woman affirmed the defining feature of paternalism, saying about her physician through a long period of her young adulthood, "I tell you he was like a father to me."[36] This doctor had also in fact been her parents' doctor, then her doctor as a child, and she had "stayed with him" into adulthood. Recalling the doctors of her youth, another octogenarian gave a brief vignette emphasizing this same point. "They came to the home with a little black bag and they sat by your bedside. And they knew your mother and father. They called them by their first name."[37] These familial and personal connections were a defining feature of this relationship from her youth. Such characterizations seem to have been about more than the presumptively benign authority of a parent, although they seem often to have reflected it. In the interviews, people broadened the sense of kinship associated with their doctors and extended it into a more complex web of personal connections.

Even the woman who remembered her doctor as having been like a father actually named that particular analogy as an element in a much more elaborate set of kinship associations. When her original doctor died, she recalled, she had continued on with the new doctor who took over his practice, as though the allegiance owed to the previous doctor was transferred with the practice. But another familial relationship also pertained. This woman explained the nature of her connection to her new doctor in the following way: "His dad had doctored my two older brothers. Then his father had been a doctor back years before. It was three generations of doctors. And my mother had gone to the older one." She knew that her doctor's professional genealogy extended back three generations, and she was able to connect the physicians in his family to her own family in all three generations. The connections between her family and her physician were like connections of kinship that extended outside the mere exchange of services and had a significance that endured through generational change.[38]

This kind of obligatory, kinship model seemed very compelling to the interviewees. In part, their relationships reflected the circumstances of rural and small-town life in the mid twentieth century. The local grocer might likewise have inherited his business and be another family's customary provider in the area. However, the ability to remember and recount particular experiences with medical care proved surprisingly powerful and detailed. In an extreme case, one informant explained with apparent conviction that she had originally been directed to her doctor by God. People rarely cited such supernatural influences. Yet they did characterize the process of getting a new physician as involving more than considered, rational self-interest. One woman, who had had to move away, leaving the practice of her long-standing physician, explained how she had found a new physician. Her previous long-term doctor "had some family" in the area to which she moved, "his mother's people." This previous doctor relied on these family connections to select a new doctor for her in the town. In this woman's recollection, the most important quality of the new doctor was that he "knew who was who." Presumably, this was a physician who would provide her with good medical care. But the dominant concern, as she recalled it, was more a matter of local connections.

Such accounts of a relationship with one physician being transferred to another physician through kinship or personal connections were common. Another eighty-two-year-old woman described how when her long-term doctor moved away, this physician had "turned me over to Dr. Walker." The relationship had seemed to be transferred by one doctor to the next. Later in the interview, this same woman noted that "these doctors retiring always make us take a choice."[39] She acknowledged a sense of choosing her next doctor but described the choice as something imposed by the previous doctor, almost as an obligation of the original relationship. These relationships to physicians had enduring qualities that were separable from the particular services rendered. When one doctor left, some of the qualities of the old relationship were preserved, carrying the memories and obligations of the previous physician, as though the relationship itself was transferred from one doctor to the next.

The process of establishing a bond with the physician was rarely described as something easy to understand or make judgments about. For people who remembered a first meeting with a doctor whom they then saw over a long period of time, we asked in the semi-structured interviews how long it was before the person knew that this doctor would be his or her doctor. Answers ranged from "right away" to "immediately" to "maybe ten minutes." One respondent put it succinctly: "Immediately. I mean if I . . . otherwise he wouldn't be it." None of the informants described a process of testing the services of these long-term physicians to decide about continuing the relationship. They did describe ruptures in previous relationships. But the relationship was primary and the services followed seemingly from the relationship. Accepting the care of these physicians was a matter of personal connection, and accepting medical services followed necessarily on accepting the relationship.

Using the interpretive framework of gift exchange and commodity exchange, we can examine the dynamics of doctor-patient relationships in different periods from Cabot's day to our own. Today, questions about consumerism in medicine generate considerable anxiety, in part because of the changing markets and mechanisms for reimbursing medical services. Can an efficient health-care business more efficiently assess and meet the needs of its consumer-patients than a traditional physician-centered organization? Should physicians fight on behalf of their dependent patients in an increasingly complex medical bureaucracy, which imposes heavy responsibilities for arranging their care on patients?

Cabot in Context

Whatever the complex sources of modern patients' activism, it reminds us that our current debates about the distinctions between patients and consumers are contingent on circumstances and historical context. At present, we fall back on this dichotomous model of the patient-consumer in many discussions about medical care. We invoke the protections of consumerism in medical decision-making, for example, through the model of informed consent. As consumers, we increasingly seek direct, individualized disclosure about the risks and benefits of our medical products, procedures, and treatments. Through reform of the medical system, we have secured somewhat improved means to act as informed consumers, better able now to collect information and to judge what seems beneficial for ourselves. At other times, however, we continue to rely heavily on expertise, turning over care rather unilaterally to a physician. A person considering a risky therapy for cancer or a specialized surgical procedure can decide to engage the services or to walk away, but may have little control over their significant content. The technical complexity of modern medicine and the dependencies of illness both incline us in this direction. We choose simply to put ourselves in the care of one physician to receive some miraculously complex package of technical services delivered with a promise of good faith. Both models of consumer and patient have application in analyzing the tasks of securing individual medical care.

Observing the institutional setting may help deepen our understanding. As in Cabot's private medical office, people will express their interests in medical services in ways that make sense in a specific institutional context. Informed consent, for example, came into use in the late twentieth century in highly specialized and compartmentalized medical institutions, where people tended to shuttle quickly through clinics, emergency departments, operating rooms, and radiology suites. Older models for certifying consent for medical services, as demonstrated among Cabot and his colleagues, relied upon personal connections with individual physicians and often involved families as surrogate decision-makers. Such fragile, personal mechanisms of consent were increasingly difficult to invoke and sustain in an institutional setting where many different physicians delivered a multitude of specialized services increasingly rapidly. Informed consent provides a simple, easily transportable mechanism for certifying consent, one that is not dependent on any one physician or family member and that moves with a patient through complex medical bureaucracies. Informed consent may on occasion be presented as the tool of activist patients who are asserting their interests against the power of biomedical mystery. Yet people have endorsed the medical model of informed consent simply because they have found themselves in institutional settings that favor its use.

The demands of patient-hood may also encourage us to resist too simple a categorization as consumers or dependent clients. A person who is sick might do best to shift readily among different strategies in seeking medical services. A sharp distinction between the dependent patient and the deliberative consumer has served certain polemical purposes, but it threatens to undermine individuals' ability to act effectively as patients. Arriving before the doctor committed to the role of the informed and judicious consumer, one might, for example, limit other potentially more effective strategies. The informed consumer takes on additional burdens of responsibility for care and may be less able to inspire a physician's personal commitment and attention. Conversely, a faithful patient cedes more to professional control and so may forfeit powers of discrimination and choice that are wisely held in reserve by the savvy consumer. Theoretical models that are attractive in describing or justifying social behaviors can limit effective improvisation in daily life.

The patients who frequented the office of Richard Cabot in the early twentieth century found a physician committed to making the benefits of scientific methods available to them in daily practice. Their experiences marked an important transition in the experience of scientific medical authority, placing them halfway between the more personal and negotiated exchanges of nineteenth-century doctors and the precise control and technical scrutiny of later twentieth-century paternalism—halfway, that is, between the hospital and the home. These patients' letters offer a privileged look at early twentieth-century medicine, balancing its technical detail against the vision that they came away with of a medicine focused on the problems of disease. Modern distinctions between the patient and the consumer of medical services seem not to distinguish among the many competing motivations that they brought with them to the medical world. In a context of numerous undifferentiated practitioners, patients sought confirmation of the value of the medical services that they had purchased. Yet they were also eager to gain the trust and the service of their physician through attentiveness to his advice and ministrations. The consumer and the dependent client were equally accessible as roles for these patients. The particular context of the practice and their individual circumstances often determined their choices in negotiating their medical care more than the structure of simple models for their interests.

NOTES

The papers of Richard Clarke Cabot [RCC] in the Harvard University Archives are cited as Cabot Papers. Individual cases from RCC's Medical Case Records are cited in the form volume:page, PR [Patient Records] (e.g., 32:118, PR). In accordance with the restrictions placed on this collection, individual patients are anonymous.

Patient Records of the Massachusetts General Hospital, Boston, are cited in the form MGH, East [or West] Medical Service, year, volume:page, PR.

ABBREVIATIONS

AIM	*Annals of Internal Medicine*
AMA	American Medical Association
BHM	*Bulletin of the History of Medicine*
BMSJ	*Boston Medical and Surgical Journal*
Countway	Francis A. Countway Library of Medicine, Boston, Rare Books and Special Collections Department
Gannet Papers	William Whitworth Gannett, manuscript notebooks, "Urine Analysis. Volume I," "Urine Analysis. Volume III," "Urine Analysis. Volume IV," and "Blood Counts [1880–85]," Francis A. Countway Library of Medicine, Rare Books and Special Collections Department
JAMA	*Journal of the American Medical Association*
MGH	Massachusetts General Hospital
NEJM	*New England Journal of Medicine*
PR	Patient Records
RCC	Richard Clarke Cabot

PREFACE

1. See Richard C. Cabot, *A Guide to the Clinical Examination of the Blood for Diagnostic Purposes* (New York: William Wood, 1896). On Cabot's early career, see Ian S. Evison, "Pragmatism and Idealism in the Professions: The Case of

Richard C. Cabot, 1868–1939" (PhD diss., University of Chicago, 1995). Also see Christopher Crenner, "Diagnosis and Authority in the Early Twentieth-Century Medical Practice of Richard C. Cabot," *BHM*, 2002, 76 (1): 30–55.

2. An insightful recent exploration of the effect of domestic practice on nineteenth-century doctoring is found in Judith Walzer Leavitt, "'A Worrying Profession': The Domestic Environment of Medical Practice in Mid-Nineteenth Century America," *BHM*, 1995, 69 (1): 1–29. A compelling examination of the sources for the ideal of genteel disinterestedness and integrity in early science appears in Steven Shapin, *A Social History of Truth: Civility and Science in Seventeenth-Century England* (Chicago: University of Chicago Press, 1994), 223–27. The relevant application of these themes to medicine in late nineteenth-century America appears, for example, in John Harley Warner, "Ideals of Science and Their Discontents in Late Nineteenth-Century American Medicine," *Isis*, 1991, 82:454–78. On the home medical office in Cabot's Back Bay neighborhood, see John D. Stoeckle, "Introduction," in *Encounters Between Patients and Doctors*, ed. id. (Cambridge, Mass.: MIT Press, 1987), 1–129, esp. 12.

3. On the novel significance of the white coat in early twentieth-century medicine in these settings, see Joseph C. Aub and Ruth K. Hapgood, *Pioneer in Modern Medicine: David Linn Edsall of Harvard* (Boston: Harvard Medical Alumni Association, 1970), 130–32.

4. Details of the social hierarchies of early twentieth-century Boston may be found in Frederic Cople Jaher, *The Urban Establishment: Upper Strata in Boston, New York, Charleston, Chicago, and Los Angeles* (Urbana: University of Illinois Press, 1982). An entertaining tour of this same terrain appears also in Cleveland Amory, *The Proper Bostonians* (New York: Dutton, 1947). Observations on the stature and composition of the medical profession in late nineteenth- and early twentieth-century in America abound. The best description of the narrowing heterogeneity of the medical profession after 1900 from the perspective of historical sociology is Paul Starr, *The Social Transformation of American Medicine* (New York: Basic Books, 1982), 30–31, 116–32, and esp. 140–44. A succinct account of the late nineteenth-century success of conventional medical practitioners to gain separate defined stature appears in William G. Rothstein, *American Physicians in the Nineteenth Century: From Sects to Science* (1972; reprint, Baltimore: Johns Hopkins University Press, 1992), 298–326. A wonderful insight into the means for distinguishing a career in late nineteenth-century American medicine can be had from Charles Rosenberg, "Making It in Urban Medicine: A Career in the Age of Scientific Medicine," in *Explaining Epidemics and Other Studies in the History of Medicine*, ed. id. (Cambridge, Mass.: Cambridge University Press, 1992), 215–42

5. For a general examination of the changing nature of medical authority, see Starr, *Social Transformation of American Medicine*, 3–29. Historians have in general emphasized the relative insecurity of the medical profession at the turn of the century, as the last two sources cited n. 4 above suggest. A useful essay characterizing the rising authority of the medical profession in the subsequent period is Allan M.

Brandt and Martha Gardner, "The Golden Age of Medicine?" in *Medicine in the Twentieth Century*, ed. Roger Cooter and John Pickstone (Amsterdam: Harwood, 2000), 21-37. For a divergent emphasis on the continuity of values and aspirations in medicine, and the solid claims that physicians maintained on the deference of their patients across the nineteenth century in the face of a changing institutional context, see Samuel Haber, *Authority and Honor in the American Professions, 1750–1900* (Chicago: University of Chicago Press, 1990), 182–86, 328–31. Writers who make an analysis of the professions similar to Haber's have still tended to emphasize the medical profession's relatively weak claims to authority into the late nineteenth century; see, e.g., Burton Bledstein, *The Culture of Professionalism: The Middle Class and the Development of Higher Education in America* (New York: Norton, 1978), 191–93.

CHAPTER 1. THE AUTHORITY OF A SCIENTIFIC DOCTOR

1. A formative early case for the determinative connection between technical change and the medical relationship is Nicholas D. Jewson, "The Disappearance of the Sick-Man from Medical Cosmology, 1770–1870," *Sociology*, 1976, 10:225–44. In *Medicine and the Reign of Technology* (New York: Cambridge University Press, 1892), Stanley Joel Reiser skillfully documents the concerns within medicine about the implications of new technology, but suggests that the impact on medical authority was somehow inherent in the techniques. Recent historiography has developed more nuanced accounts, but the older versions remain quite influential, especially in writings from bioethics. See the introductory section of Chapter 7 for a brief review.

2. The examination of medical authority from historical sociology by Paul Starr, *The Social Transformation of American Medicine* (New York: Basic Books, 1982), 3–29, is helpful here but looks at the question from the other side. Starr examines the growth of collective professional authority and speculates about its implications for individual interactions between physician and patient. I examine individual interactions as they were characterized by physician and patient and set them in the local, institutional context of a private clinic. My account is closer to the version of events given in Edward Shorter, *Bedside Manners: The Troubled History of Doctors and Patients* (New York: Simon & Schuster, 1985), although with considerably more to say about technical practice. I have relied in several places on the discussion of the relationship between expertise and professional authority in Anthony Giddens, *The Consequences of Modernity* (Stanford, Calif.: Stanford University Press, 1990), 79–92. This latter work introduces the useful notion of face-to-face contact with the expert as a critical node in modern technical systems. In trying to distinguish between individual influence and general professional power, I have drawn also from the broad analysis developed in Richard Sennett, *Authority* (New York: Norton, 1980), 52–121, especially concerning distinctions between relations to paternalism and to client autonomy. Erving Goffman, *The Presentation of Self in*

Everyday Life (Garden City, N.Y.: Doubleday Anchor, 1959) provides additional useful analytic tools for a sociological analysis of individual authority and introduces many of the concepts that Giddens scrutinizes.

3. See Malcolm Nicolson, "The Art of Diagnosis and the Five Senses," in *Companion Encyclopedia of the History of Medicine*, ed. William F. Bynum and Roy Porter (London: Routledge, 1993), 801–25. Reiser, *Medicine and the Reign of Technology*, 231, indirectly suggests this characteristic of medical diagnosis but does not attempt to resolve it with his overarching critique of "the gap that technology has helped to create between the patient and physician."

4. Erving Goffman, *Asylums: Essays on the Social Situation of Mental Patients and Other Inmates* (New York: Anchor Books, 1961), esp. 12–74.

5. MGH, East Medical Service, 1903, 581:120, PR. On the growth of professional nursing around new medical techniques, see esp. Margarete Sandelowski, *Devices and Desires: Gender, Technology, and American Nursing* (Chapel Hill: University of North Carolina Press, 2000).

6. See the insight on the shifting place of the private medical office in David Rosner, *A Once Charitable Enterprise: Hospitals and Health Care in Brooklyn and New York, 1885–1915* (New York: Cambridge University Press, 1982), 31–35. By 1900, the changes that brought the private practice of medicine routinely into American hospitals had just begun: see on this, e.g., Charles E. Rosenberg, *The Care of Strangers: The Rise of America's Hospital System* (New York: Basic Books, 1987), 244–58. For medical education in the period, see Kenneth Ludmerer, *Learning to Heal: The Development of American Medical Education* (New York: Basic Books, 1985). For the development of the research enterprise, see, e.g., Harry M. Marks, *The Progress of Experiment: Science and Therapeutic Reform in the United States, 1900–1990* (New York: Cambridge University Press, 1997).

7. The historical development of the medical office has attracted relatively little attention. The economic dimensions of private practice have occasionally come under scrutiny by historians, as in Donald E. Konold, *A History of American Medical Ethics, 1847–1912* (Madison: State Historical Society of Wisconsin, 1962). For a focused analysis, see Thomas Goebel, "American Medicine and the 'Organizational Synthesis': Chicago Physicians and the Business of Medicine, 1900–1920," *BHM*, 1994, 68:639–63. See also George Rosen, *Fees and Fee Bills: Some Economic Aspects of Medical Practice in Nineteenth Century America* (Baltimore: Johns Hopkins Press, 1946). One good start on understanding the private medical office is John D. Stoeckle, "Introduction," in *Encounters Between Patients and Doctors*, ed. id. (Cambridge, Mass.: MIT Press, 1987), 1–129. Discussion of changes in the economic circumstances of the medical profession and their repercussions for reform in early twentieth-century medicine appears in James G. Burrow, *Organized Medicine in the Progressive Era: The Move Toward Monopoly* (Baltimore: Johns Hopkins University Press, 1977), 105–18. A careful history of the organization of private medical practice in England appears in Anne Digby, *Making a Medical Living: Doctors and Patients in the English Market for Medicine, 1720–1911* (New York: Cambridge University Press,

1994), esp. 197–253; see also id., *The Evolution of British General Practice, 1850–1948* (New York: Oxford University Press, 1999), 93–186, 224–55.

8. For a helpful review and critique of the American physician's established defense of the doctor-patient relationship in private practice, see Herman Miles Somers and Anne Ramsay Somers, *Doctors, Patients, and Health Insurance: The Organization and Financing of Medical Care* (Washington, D.C.: Brookings Institution, 1961 [repr. 1966]), 455–75.

9. This distinction between backstage and onstage performance of expertise derives from the classic discussion in Goffman, *Presentation of Self,* 170–75.

10. Memorial Seminar transcript, 18 Jan. 1958, Cabot Papers, fld. "Cabot Seminar Papers," box "Memorial Seminar."

11. Starr, *Social Transformation,* 140–44.

12. The course of this change is implied in the details of Rosenberg, *Care of Strangers,* although he refrains from talking about physician-patient relationships.

13. Ibid., 184–89. See also Starr, *Social Transformation,* 145–79. Perspective on the growth of the American hospital system emphasizing the inherent commercial incentive appears in Rosemary Stevens, *In Sickness and in Wealth: American Hospitals in the Twentieth Century* (1989; reprint, Baltimore: Johns Hopkins University Press, 1999), 30–39, 111–14.

14. MGH, East Medical Service, 1897, 487:158, 40, PR, Countway.

15. A review of the many significances of science for medicine can be found in John Harley Warner, "Science in Medicine," in *Historical Writing on American Science: Perspectives and Prospects,* ed. Sally Gregory Kohlstedt and Margaret W. Rossiter (Baltimore: Johns Hopkins University Press, 1985), 37–58. "Technological medicine" comes rather closer to describing what I have in mind but lacks any solid connection to the language of the times. For a view of these practices from the perspective of their technologies, see Joel Howell, *Technology in the Hospital: Transforming Patient Care in the Early Twentieth Century* (Baltimore: Johns Hopkins Press, 1996), who along with Reiser has principally examined diagnostic technology, documenting its spreading use and contested acceptance by physicians. For a focus on interaction between new diagnostic technology and disease definition, see Keith Wailoo, *Drawing Blood: Technology and Disease in Twentieth Century America* (Baltimore: Johns Hopkins University Press, 1997).

16. The argument that the disease model is a fundamental organizational feature of twentieth-century medicine may be found in Charles E. Rosenberg, "The Tyranny of Diagnosis: Specific Entities and Individual Experience," *Milbank Quarterly,* 2002, 80 (2): 237–60. A recent approach to studying changes in popular perceptions is exemplified in Bert Hansen, "New Images of a New Medicine: Visual Evidence for the Widespread Popularity of Therapeutic Discoveries in America after 1885," *BHM,* 1999, 73 (4): 629–78; and "America's First Medical Breakthrough: How Popular Excitement about a French Rabies Cure in 1885 Raised New Expectations for Medical Progress," *American Historical Review,* 1998, 103 (2): 373–418.

17. Letter to RCC, 7 Nov. 1916, Medical Case Records, Cabot Papers, 32:118, PR.

18. RCC, letter to Dr. G. S. Foster, 4 Oct. 1912, 18:6, PR.

19. For the present day disease-illness distinction, see Arthur Kleinman, Leon Eisenberg, and Byron Good, "Culture, Illness, and Care: Clinical Lessons from Anthropologic and Cross-Cultural Research," *AIM*, 1978, 88 (2): 251–58, which is often cited for its cogent statement of the modern significance. For a passionate analysis of the implications of this distinction, see Eric Cassell, *The Nature of Suffering and the Goals of Medicine* (New York: Oxford University Press, 1991). Also see Arthur Kleinman, *The Illness Narratives* (New York: Basic Books, 1988), esp. 3–30. These authors treat the distinction between the physician's understanding of disease and the patient's experience of illness as a special characteristic of late twentieth-century technical medicine.

20. Still the best source for the history of these concepts is Owsei Temkin, "The Scientific Approach to Disease: Specific Entity and Individual Sickness," in id., *The Double Face of Janus and Other Essays in the History of Medicine* (Baltimore: Johns Hopkins University Press, 1977), 441–55. The theme has been reviewed and expanded recently by Charles E. Rosenberg, "What Is Disease? In Memory of Owsei Temkin," *BHM*, 2003, 77 (3): 491–505. For examples of the issue in diagnostic practice in an earlier period, see Nicolson, "Art of Diagnosis," 801–25. For the reference to the *Hippocratic corpus*, see Helen King, *Hippocrates' Woman: Reading the Female Body in Ancient Greece* (New York: Routledge, 1998), 31–35.

21. A version of the turn-of-the-century debate appears in James Jackson Putnam, "Not the Disease Only but Also the Man," *BMSJ*, 1899, 141:53–57. Cf. S. C. Gordon, "Common Sense in Medicine," *BMSJ*, 1891, 25:233–36, 261–63. As examined further in Chapter 3, a similar theme appears in Alfred Worcester, "Past and Present Methods in the Practice of Medicine," *BMSJ*, 1912, 166:159–64.

22. For a formative discussion of the illness-disease distinction in the early twentieth-century, see G. Canby Robinson, *The Patient as a Person: A Study of the Social Aspects of Illness* (New York: Commonwealth Fund, 1939), 3–5. And see earlier Austin Fox Riggs, "The Significance of Illness," in *Physician and Patient: Personal Care*, ed. L. Eugene Emerson (Cambridge, Mass.: Harvard University Press, 1929), 100–21. Similar distinctions were maintained implicitly in the other essays collected in Emerson, ibid. Unlike Kleinman, Cassell, and the commentators cited in n. 19 above, these early twentieth-century essayists allowed that illness might also be a perspective to be adopted by the physician, rather than solely a name for the patient's independent perspective.

23. A wonderful treatment of this concern as a historical problem can be found in Georges Canguilhem, *The Normal and the Pathological*, trans. Carolyn R. Fawcett and Robert S. Cohen (New York: Zone Books, 1989). Canguilhem identifies neatly, by reference to the writings of René Leriche in the 1930s, the parallels between the two distinctions of functional and organic and of illness and disease: ibid., 46–53.

24. A clear statement from the time on the significance of functional diseases occurs in James Jackson Putnam, "The Value of the Physiological Principle in the Study of Neurology," *BMSJ*, 1904, 151 (December): 641–47. This distinction between functional and organic disease found wide discussion in the nineteenth century. But it held a special relevance to neurology in the day, as, e.g., in Siegfried Block, "Hysteria Versus Neurasthenia," *Medical Record (NY)*, 1918, 94 (8): 326–28. Recent secondary analyses have also tackled the functional and structural distinction as physicians applied it to evolving ideas about heart disease: Christopher Lawrence, " 'Definite and Material': Coronary Thrombosis and Cardiologists in the 1920s," in *Framing Disease: Studies in Cultural History*, ed. Charles E. Rosenberg and Janet Golden (New Brunswick, N.J.: Rutgers University Press, 1992), 51–82. Also see Robert Alan Aronowitz, *Making Sense of Illness: Science, Society and Disease* (New York: Cambridge University Press, 1998), 84–110, on functional and structural debates in cardiology.

25. John C. Hemmeter, "Science and Art in Medicine: Their Influence in the Development of Medical Thinking," *JAMA*, 1906, 46:243–48. Cf. Theodore C. Janeway, "Some Sources of Error in Laboratory Clinical Diagnosis," *Medical News*, 1901, 78:700–706. Also see Alfred Loomis, "President's Address to the Association of American Physicians," *BMSJ*, 1893, 129:144–45. For an interesting connection made in the day between the science of medicine and the distinction between functional and organic disease, see Samuel J. Meltzer, "Chairman's Address, Section of Physiology," *Congress of Arts and Sciences. Universal Exposition*, 1906, 5:395–402. A historian's approach to this issue is found in Lester S. King, *Transformations in American Medicine: From Benjamin Rush to William Osler* (Baltimore: Johns Hopkins University Press, 1991), 226–28.

26. The classic discussion of how medical judgment figured into the establishment of scientific medical knowledge appeared just a decade after Cabot's heyday in Ludwig Fleck, *Entstehung und Entwicklung einer wissenschaftlichen Tatsache: Einführung in die Lehre vom Denkstil und Denkkollectiv* (Basel: B. Schwabe, 1935), trans. Fred Bradley and Thaddeus J. Trenn under the title *Genesis and Development of a Scientific Fact*, ed. Thaddeus J. Trenn and Robert K. Merton (Chicago: University of Chicago Press, 1979).

27. For further biographical details, see Paul Dudley White, "Obituary, Richard Clarke Cabot, 1863–1939," *NEJM*, 1939, 220, 1049–52, and Robert Schulyer, ed., "Richard Clarke Cabot," *Dictionary of American Biography, suppl. 2* (New York: Scribner, 1958), 83–85. Ian S. Evison, "Pragmatism and Idealism in the Professions: The Case of Richard C. Cabot, 1868–1939" (PhD diss., University of Chicago, 1995) includes a comprehensive bibliography. For the dates and offices, see Harvard University, *Harvard University Quinquennial Catalogue of the Officers and Graduates, 1636-1930* (Cambridge: University Press, 1930), 55, 60, 346, 336; Frederic Washburn, *The Massachusetts General Hospital: Its Development, 1900–1935* (Boston: Houghton Mifflin, 1939), 86–87, 115–16, 126, 433, 435.

28. Richard C. Cabot, *A Guide to the Clinical Examination of the Blood for Diagnostic Purposes* (New York: William Wood, 1896).

29. RCC, fld. "Autobiographical Notes," p. 7, box 109, Biographical Materials, Cabot Papers. See also Richard C. Cabot, "The 'Bay State's' First Trip to Porta Rico" [sic], *BMSJ*, 1898, 139 (15): 378–9. Further information on this military medical endeavor is found in Vincent J. Cirillo, *Bullets and Bacilli: The Spanish-American War and Military Medicine* (New Brunswick, N.J.: Rutgers University Press, 2004), 16, 27, 42.

30. On Cabot's reputation in his day as a critic of conventional practice, see, e.g., Arthur C. Jacobson, "Some Thoughts on the Geography of Medical Greatness," *Medical Times*, January 1915, 24–25. See Richard C. Cabot, "The Ideal of Accuracy in Clinical Work: Its Importance, Its Limitation," *California State Journal of Medicine*, 1904, 2 (12): 361–63. Cf. Richard C. Cabot, "Limitations of Urinary Diagnosis," *Johns Hopkins Hospital Bulletin*, 1904, 15 (158): 174–77. Also see Richard C. Cabot, "Better Doctoring for Less Money," *American Magazine*, 1916, 81 (4): 7–9, 77–78.

31. White, "Obituary," 1051.

32. Arthur K. Shapiro and Elaine Shapiro, *The Powerful Placebo: From Ancient Priest to Modern Physician* (Baltimore: Johns Hopkins University Press, 1997), 175. See also Howard M. Spiro, *Doctors, Patients, and Placebos* (New Haven, Conn.: Yale University Press, 1986), 121–23. Cf. Howard Brody, *Placebos and the Philosophy of Medicine: Clinical, Conceptual, and Ethical Issues* (Chicago: University of Chicago Press, 1980).

33. RCC, letter to Ella Lyman, 11 June 1893, fld. "Ella Lyman II," box 14, Correspondence, Cabot Papers.

34. RCC, "Autobiographical Notes," 8–9.

35. Aub and Hapgood, *Pioneer in Modern Medicine*, 60–65, 138–39. See also Thomas A. Dodd, "Opening Windows: Richard Cabot and the Care of the Patient During America's Progressive Era, 1890–1920" (undergraduate honors thesis, Harvard University, Department of History of Science, June 1980). Writing to his wife many years later, Cabot still regretted this missed promotion, saying that even she "didn't know how hard that hit." RCC, letter to Ella Lyman Cabot, 26 Feb. 1931, fld. 7, box 14, Cabot Papers.

36. RCC, "Autobiographical Notes," 9.

37. For a partial exception, see RCC's brief quotation condemning the mercy killing of an "unfit" infant, "Defective Baby Is Dead," *Boston Daily Globe*, 18 Nov, 1915, 5.

38. "Notes of APM Discussion with RCC 1936," box 18, Research Materials of Ada P. McCormick, Cabot Papers.

39. The physician and ethicist Howard Spiro insightfully identifies this question about Cabot, although he had only seen indirect evidence of Cabot's involvement in his brother's death: Spiro, *Doctors, Patients, and Placebos*, 121–23.

40. Thirty-six bound volumes with two packets of unbound leaves of Cabot's

manuscript private patient records were archived as Richard C. Cabot, Medical Case Records, volumes I–XXVI and two folders, 1896–1926, Cabot Papers.

41. McCormick's research for a biography of Cabot (25 containers) is included in Biographical Materials, Cabot Papers.

42. See, e.g., the astute examination of the experiences of Deborah Final Fiske in Sheila Rothman, *Living in the Shadow of Death: Tuberculosis and the Social Experience of Illness in American History* (Baltimore: Johns Hopkins University Press, 1994), 77–127. Parallel trends in historiography are helping to capture a more inclusive view of the experience of illness and healing, as in the recent scholarship of Emily K. Abel, *Hearts of Wisdom: American Women Caring for Kin, 1850–1940* (Cambridge, Mass.: Harvard University Press, 2000). The literature on the history of disability represents a further development in this line of inquiry: Catherine J. Kudlick, "Disability History: Why We Need Another 'Other,' " *American Historical Review*, 2003, 108 (June): 763–93. One programmatic inspiration for this history of the patient appears in Roy Porter, "The Patient's View: Doing Medical History from Below," *Theory and Society*, 1987, 14:167–74. A general model for this approach can be found, e.g., in Dorothy Porter and Roy Porter, *Patient's Progress: Doctors and Doctoring in Eighteenth-Century England* (Oxford: Polity Press, 1989), 155–63.

43. In June 1908, Cabot made a large mailing of a form to his patients requesting information on "the outcome of the illness for which I saw you"; see an example of the form letter, 2:58, PR. In this respect, Cabot's correspondence bears a close relationship to the papers of Dr. Elliott Joslin, recently examined in Chris Feudtner, *Bittersweet: Diabetes, Insulin, and the Transformation of Illness* (Chapel Hill: University of North Carolina Press, 2003).

44. See for some of the more pointed examples: letter to RCC, June 1908, 2:58, PR; letter to RCC, 24 May 1913, 25:40, PR; letter to RCC, June 1908, 1:149, PR; letter to RCC, 11 Feb. 1913, comparing Dr. [F?]owle, letter to RCC, 1 Apr. 1912, 18:99, PR; letter to RCC, 10 Mar. 1907, 6:37, PR; letter to RCC, 31 Dec. 1914, 1:65, PR; letter to RCC, 1 Aug. 1899, 1:31, PR; letter to RCC, 14 Jan. 1912, 15:18, PR; letter to RCC, June 1908, 2:261; letter to RCC, 30 Jan. 1904, 2:222, PR; letter to RCC, 10 June 1908, 6:8; letter to RCC, n.d., 9:121, PR; letter to RCC, June 1908, 2:209; letter to RCC, 19 Oct. 1900, 1:143, PR; letter to RCC, 25 Mar. 1902, 2:55; letter to RCC, 24 Jan. 1911, 18:59, PR; letter to RCC, 19 May 1902, 2:77, PR; and letter to RCC, 13 Nov. 1915, 23:62, PR.

45. Letter to RCC, 31 Oct. 1903, 2:235, PR; and entry 20 July 1903, 2:235, PR.

46. EG, letter to RCC, 31 May [1903?], fld. "G," Personal Correspondence, Cabot Papers; and letters to RCC seeking advice rather than an appointment or care, 7 and 22 Nov. 1916, 32:106, PR.

47. Porter and Porter, *Patient's Progress*, 89, 139, 208–9.

48. Letter to RCC, 16 Oct. 1912; and Agnes C. Victor, MD, letter to RCC, 9 Oct. 1912, 23:40, PR.

49. Thomas McGill and Mayur M. Desai, "Restricted Activity among Com-

munity-Living Older Persons: Incidence, Precipitants and Health Care Utilization," *AIM*, 2001, 135 (5): 313–19.

50. Once again the work of Anthony Giddens provides a valuable theoretical basis for understanding the modern interactions of client and expert: "Modern social life is a complex affair, and there are many 'filter-back' processes whereby technical knowledge, in one shape or another, is reappropriated by lay persons and routinely applied in the course of their day-to-day activities. As was mentioned earlier, the interaction between expertise and reappropriation is strongly influenced, among other things, by experiences at access points [i.e., face-to-face contact with experts]," *Consequences of Modernity*, 145. I recognize also the influence of reading Michel Foucault, *Birth of the Clinic: An Archeology of Medical Perception*, trans. A. M. Sheridan Smith (New York: Pantheon Books, 1973) at a formative stage in this project, but when I went back to try to undo the damage, it seemed only to increase.

51. Flipping through the index, I estimated the number of multiple-volume cases at about 1 in 70.

52. Matthew Mayo, director, Medical Statistics and Research Design Unit, University of Kansas Medical Center, helped with the application of this proportional random sampling method. He has my thanks and my added acknowledgement that any slip in the use of these techniques is my responsibility.

53. Entries, 27 Mar. 1909, 12:92; 31 July 1900, 1:124; 3 Jan. 1900, 1:76; 2 July 1902, 2:122, PR.

54. Entry, 12 Mar. 1912, 22:19, PR.

55. Joan Jacobs Brumberg, " 'Something Happens to Girls': Menarche and the Emergence of the Modern American Hygienic Imperative," *Journal of the History of Sexuality*, 1993, 4:99–127. Cf. the way that social identity mattered intimately to Deborah Vinal Fiske in interpreting her need for medical attention to her consumption: Rothman, *Living in the Shadow of Death*, 77–127.

56. See on professional authority in the Progressive Era for medicine specifically Robert H. Wiebe, *The Search for Order, 1877–1920* (New York: Oxford University Press, 1967), 114–16.

Chapter 2. Organizing a Private Office

1. Entries, 17 Jan. 1902, 2:26; 3 Mar. 1902, 2:47; 20 Oct. 1902, 2:130, PR, Cabot Papers.

2. For a sense of the domestic context of medical practice in the nineteenth century, see Judith Walzer Leavitt, " 'A Worrying Profession': The Domestic Environment of Medical Practice in Mid-Nineteenth Century America," *BHM*, 1995, 69:1–29.

3. Cabot, "Autobiographical Notes," pp. 6–7, box 109, Cabot Papers; Richard C. Cabot, "Leukocytosis as an Element in the Prognosis of Pneumonia," *BMSJ*, 1893, 129 (5): 117–8. The work on blood analysis appeared in Richard C. Cabot,

"Acute Leukemia," *BMSJ*, 1894, 130 (21): 507–11. Also appearing the same year was Richard C. Cabot, "The Diagnostic and Prognostic Importance of Leucocytosis," *BMSJ*, 1894, 130 (2): 277–82. Cabot's first book was *A Guide to the Clinical Examination of the Blood for Diagnostic Purposes* (New York: William Wood, 1896).

4. Dr. A.E.H., letter to RCC, 12 Mar. 1902, 2:52, PR.

5. Dr. Frank Day, letter to RCC, 8 Oct. 1899, 1:185, PR.

6. Letter to RCC, 12 Mar. 1902, 2:52, PR; letter to RCC, June 1908, 2:58, PR; letter to RCC, n.d. [ca. December 1900], 1:184, PR.

7. John W. Farlow, "Obituary: William Whitworth Gannett, 1853–1929," *NEJM*, 1929, 200 (22): 1177–80, Gannett Papers.

8. Gannett, "Volume I," pp. 143, 267, 155, 267, 15; entry dated 11 May 1889, "Volume IV," n.p., Gannett Papers.

9. F. G. Morrill, letter to Gannett, 2 Aug. 1886, "Volume II," n.p.; and M. Holbrook, letter to Gannett, 4 Jan. 1888, "Volume III," p. 61, Gannett Papers.

10. Advertisement, *Charlotte Medical Journal*, 1901, 19:xxiv. For similar advertisements, see, e.g., *Cleveland Medical Gazette*, 1901, 17:63; *Medical Visitor* (Chicago), 1902, 18:n.p.

11. Advertisement, *Clinical Reporter* (St Louis), 1905, 18:396.

12. Letters to Gannett, 25 Sept. 1885, "Volume I," p. 31; 2 Sept. 1885, "Volume I," p. 20; 20 Feb. 1888, "Volume III," p. 94; entry, 21 July 1888, "Volume III," p. 258, Gannett Papers.

13. C. B. Porter, letter to Gannett, 3 May 1889, "Volume IV," n.p.; Gannett, letter, n.d., and entry, 7 May 1889, "Volume IV," n.p.; entry, 19 Dec. 1885, "Volume I," p. 89; entry, 1 Aug. 1887, "Volume II," p. 211; letter to Gannett and entry, 23 July 1885, "Volume I," p. 12, Gannett Papers.

14. Farlow, "Obituary: William Whitworth Gannett," 1178.

15. James Lewis, letter to A. N. Blodgett, 13 Apr. 1892, Papers of A. N. Blodgett, Countway.

16. On fees for laboratories tests, see letter to RCC, 12 May 1902, PR 2:52. These records provide a wealth of information on the location and nature of local medical practices. See on syphilis testing F. L. Burnett, letter to RCC, 5 Mar. 1911, PR, 18:112; entry, 22 June 1912, PR, 23:8. Additional information on laboratories, practices, and individual addresses from the period were found in AMA, *American Medical Directory*, vol. 1 (Chicago: American Medical Association Press, 1906). See Boston Medical Publishing, *Medical Directory of Greater Boston* (Boston: Boston Medical Publishing Co., 1906). See also Richard R. Pettigrew, *Pettigrew's New England Professional Directory* (Boston: Garden Press, 1904). Locations of laboratories may be found in *Polk's Medical Register and Directory of North America* (Detroit: R. L. Polk, 1913).

17. See for examples of use of equipment: 1:79; 2:85; 2:89; 1:135; 2:103, PR. On the hemaglobinometer, see M. L. Verso, "The Evolution of Blood-Counting Techniques," *Medical History*, 1964, 8:149–58.

18. Parallel changes in the relationships between medical offices and hospitals

were noted by David Rosner investigating the growth of private medical practice in New York City at the turn of the century. See David Rosner, *A Once Charitable Enterprise: Hospitals and Health Care in Brooklyn and New York, 1885–1915* (New York: Cambridge University Press, 1982), 31– 35.

19. Henry A. Christian, "The History of Medicine in Boston, 1880–1930," in *Fifty Years of Boston: A Memorial Volume Issued in Commemoration of the Tercentenary of 1930*, ed. Elisabeth M. Herlihy (Boston: Tercentenary Committee, 1932), 417–34. For the development of Back Bay, see also Frederic Cople Jaher, *The Urban Establishment: Upper Strata in Boston, New York, Charleston, Chicago, and Los* Angeles (Urbana: University of Illinois Press, 1982), 25, 56. Also see Edith Guerrier, "Incidents in the Life of Boston, 1880–1930," in *Fifty Years of Boston, ed.* Herlihy, 715–50; and Sam Bass Warner Jr., *Streetcar Suburbs: The Process of Growth in Boston, 1870–1900*, 2d ed. (Cambridge, Mass.: Harvard University Press, 1978), 2, 7, 17, 32, 38.

20. These addresses can be readily sorted out by comparing the entries in the professional directories cited in n. 16 above.

21. Henry K. Beecher and Mark D. Altschule, *Medicine at Harvard: The First 300 Years* (Hanover, N.H.: University Press of New England, 1977), 461–74.

22. See directories cited in n. 16 above.

23. Oliver Wendell Holmes, "Homeopathy and Its Kindred Delusions," in id., *Medical Essays: 1842–1882* (Boston: Houghton, Mifflin, 1899), 1–27. On the reformation of the American medical profession, see Ronald L. Numbers, "The Fall and Rise of the American Medical Profession," in *Sickness and Health in America: Readings in the History of Medicine and Public Health*, ed. Judith Walzer Leavitt and Ronald L. Numbers (Madison: University of Wisconsin Press, 1985), 185–96. For analysis more sympathetic to the physician, see Kenneth M. Ludmerer, *Learning to Heal: The Development of American Medical Education* (New York: Basic Books, 1985), 166–90. Also see Paul Starr, *The Social Transformation of American Medicine* (New York: Basic Books, 1982), 79–144, and William G. Rothstein, *American Physicians in the Nineteenth Century: From Sects to Science* (Baltimore: Johns Hopkins University Press, 1972). For medical schools and matriculants, see AMA, *American Medical Directory*. Cf. also Steven J. Peitzman, *A New and Untried Course: Woman's Medical College and Medical College of Pennsylvania, 1850–1998* (New Brunswick, N.J.: Rutgers University Press, 2000). See also Ludmerer, *Learning to Heal*. And see Regina Morantz-Sanchez, *Sympathy and Science: Women Physicians in American Medicine* (New York: Oxford University Press, 1985).

24. Letter to RCC, 7 May 1902, 1:168, PR. Entry, 13 Mar. 1908, 9:137; letter to RCC, 25 Jan. 1912, 18:92; entry, 26 Oct. 1916, 32:9; letter to RCC, 7 May 1902, 1:168; and letter to RCC, n.d., 15:47, PR. It is useful to refer to the concept of a medical marketplace illustrated in an earlier era: Dorothy Porter and Roy Porter, *Patient's Progress: Doctors and Doctoring in Eighteenth-Century England* (Oxford: Polity Press, 1989), 208–9.

25. Entries at 10:147, PR.

26. These practitioners can be traced through professional directories and phone directories that variously give both professional office and home addresses. See New England Telephone, *The Boston Directory* (Boston: Sampson & Murdock, 1904); Pettigrew, *Pettigrew's New England Professional Directory*. For Brainerd, see Lionel Topaz, *The Red Book of Eye, Ear, Nose and Throat Specialists*, 1st ed. (Chicago: Lionel Topaz, 1915). And see also directories cited in n. 16 above.

27. For the nineteenth-century significance of the private medical office, see Charles Rosenberg, "The Practice of Medicine in New York a Century Ago," in *Explaining Epidemics and Other Studies in the History of Medicine* (Cambridge: Cambridge University Press, 1992), 131–33. By 1928, office visits outnumbered home visits in physicians' practices nationally; and by the 1950s, home visits were the smallest single element among physicians' contacts with patients. The value attached to home visits seems to have risen as fast as their practical significance declined, making it difficult to get a clear fix on their role in medical practice at any given moment. See Isidore S. Falk, Margaret C. Klem, and Nathan Sinai, *The Incidence of Illness and the Receipt and Costs of Medical Care Among Representative Family Groups*, Publications of the Committee on the Costs of Medical Care (Chicago: University of Chicago Press, 1933), 283. The complete statistics are noted in U.S. National Center for Health Statistics, *Physician Visits—Volume and Interval since Last Visit: U.S. 1971, Public Health Service*, ser. 10 (1975), 5, 9. For advice on home visits, see Daniel W. Cathell, *The Physician Himself and Things That Concern His Reputation and Success*, 10th ed. (Philadelphia: F. A. Davis, 1900), 20–1, 24, 120, 166–67. A similar account appears in Arpad G. Gerster, *Recollections of a New York Surgeon* (New York: Paul B. Hoeber, 1917), 169–70.

28. Letter to RCC, 24 June 1911, 15:4, PR; letter to RCC, 13 Apr. 1914, 27:84, PR; cf. similarly letter to RCC, 22 Feb. 1915, 22:9, PR.

29. Dr. Daniel R. Brown, letter to RCC, 25 May 1902, 2:85, PR.

30. Letter to RCC, 20 Mar. 1906, 3:97, PR.

31. See n. 16 above for professional directories of office locations.

32. John D. Stoeckle, "Introduction," in *Encounters Between Patients and Doctors: An Anthology*, ed. id. (Cambridge, Mass.: MIT Press, 1987), 1–129.

33. Roger I. Lee, *The Happy Life of a Doctor* (Boston: Little, Brown, 1956), 33–34.

34. See the directories cited in n. 16 above.

35. Joseph McDowell Mathews, *How to Succeed in the Practice of Medicine* (Philadelphia: W. B. Saunders, 1905), 32–41. Also see Cathell, *Physician Himself*, 5–10.

36. Cathell, *Physician Himself*, 7.

37. Ibid.

38. Richard C. Cabot, "Better Doctoring for Less Money," *American Magazine*, 1916, 81 (4): 7–9, 77–78.

39. John Harley Warner, "Ideals of Science and Their Discontents in Late Nineteenth-Century American Medicine," *Isis*, 1991, 82:454–78.

40. The records used a system of designations to track the physicians who shared patients with Cabot. The system distinguished between patients referred to Cabot and patients seen together in consultation. Cabot was assiduous about identifying and communicating with the referring and consulting physicians. Patients with no notations about referring or consulting physicians must have made their way on their own, or concealed or lost track of their original physicians. There is evidence of slippage in the record-keeping, but the system seems to have been generally thorough and reliable. Cabot used the notation "seen c[um] [with]" the consulting physician, to indicate that he was present with the doctor. These records were often pasted into the record books on separate sheets of paper, showing that they were written outside the office. One possible source of bias was the method of recording new cases. Each new record consisted of a first new visit with subsequent follow-up visits recorded with the original case. If there was a gap of years between the original visit and subsequent visits, however, the person might appear to be self-referred, when in fact he or she had originally been sent by another physician.

41. Dr. Perley, letter to RCC, 9 Nov. 1913, 25:142, PR.

42. Letter to RCC, 1 Feb. 1914, 25:40, PR; letter to RCC, n.d., 5:145, PR.

43. Gerster, *Recollections of a New York Surgeon*, 192.

44. Warner, "Ideals of Science and Their Discontents," passim.

45. Sharon R. Kaufman, *The Healer's Tale: Transforming Medicine and Culture* (Madison: University of Wisconsin Press, 1993), 156–57, 234, 306–7.

46. Lewis J. Moorman, *Pioneer Doctor* (Norman: University of Oklahoma Press, 1951), 77, 131, 135–36.

47. AMA, "The Code of Ethics," *JAMA*, 1892, 19 (21): 605–12. Cf. AMA, "Principles of Medical Ethics," *JAMA*, 1903, 40 (20): 1371, 1379–81. The quotation is from AMA, "Principles of Medical Ethics: A Proposed Revision," *JAMA*, 1912, 54 (23): 1790–93.

48. Berthold Ernest Hadra, *The Public and the Doctor* (Dallas: Franklin Press, 1902) 103. See a similar discussion in Mathews, *How to Succeed*, 85–92.

49. Editorial, "Commissions," *JAMA*, 1899, 33 (19): 1173–74.

50. Donald E. Konold, *A History of American Medical Ethics, 1847–1912* (Madison: State Historical Society of Wisconsin, 1962), 68–71 and passim. See also a lively discussion of fee-splitting in Hugh Cabot, *The Doctor's Bill* (New York: Columbia University Press, 1935), 125–28.

51. Letter to RCC, 8 Apr. 1914, 27:84, PR; letter to RCC, 13 July 1910, 16:138; letter to RCC, n.d., 5:235, PR.

52. Pettigrew, *Pettigrew's Directory*, 233.

53. Entry, 29 June 1912; entry, 14 Oct. 1912; letter to RCC, 11 Sept. 1912, 23:27, PR.

54. Walter Sawyer, letter to RCC, 4 May 1910, 16:43, PR.

55. Ibid.

56. Letter to J. J. Putnam, 25 Sept. 1913, 25:145, PR.

57. Letter to RCC, 9 Mar. 1914, 27:60, PR.

58. Sarah Bond, MD, letter to RCC, n.d., 4:176, PR.

59. Letter to RCC, 6 Nov. 1906, 5:171; letter to RCC, 21 June 1908, 6:244, PR.

60. Letter to RCC, n.d., 6:239; letter to RCC, n.d., 6:89, PR. See similarly letter to RCC, 21 Nov. 1911, 21:14, PR.

61. Letter to RCC, 12 Mar. 1915, 22:51; entry, 16 Nov. 1908, 11:38; letter to RCC, 22 Sept. 1905, 5:176; entry, 23 May 1897, fld. "1896–1897," PR. Cabot did report working with homeopathic doctors also in consultation: Richard C. Cabot, *Differential Diagnosis: Presented Through an Analysis of 385 Cases*, 2d ed. (Philadelphia: W. B. Saunders, 1912) 400.

62. Entries in 1:199, 2:26, 2:47, 3:254, 4:34, 5:4, 13:106, 26:37, PR.

63. Letter to RCC, 20 Mar. 1906, 3:97, PR.

64. Letter to RCC, 29 Nov. 1909, 14:100, PR.

65. Letter to RCC, 12 Mar. 1912, 22:18, PR.

66. Letter to RCC, 10 Feb. 1914, 26:37, PR. Letter to RCC, 12 Mar. 1912, 22:18, PR.

67. Letter to RCC, 25 Mar. 1914, 27:75, PR.

68. Letter to RCC, 7 Nov. 1901, 1:3, PR.

69. Entries, 9 through 30 July 1897, fld. "1896–1897," PR.

70. Entry, 8 May 1905, 4:45, PR.

71. Because of agreements about confidentiality, I have chosen not to pursue the identity of individual patients by tracing them outside of the records contained in Cabot's files.

72. Entries in 2:97; 2:85; 2:155; 2:116; 2:210; 2:113; 2:96, 3:56, PR.

73. Entries, 13 Jan. 1908, 8:23; 25 July 1902, 2:102, PR.

74. Entry, 12 Mar. 1912, 22:19; entry, 26 May 1900, 1:109; entry, 29 May 1912, 22:124; entries, 15 and 27 May 1900, 1:104; entry, 16 Oct. 1900, 1:155; entry, 5 Mar. 1909, 12:40; 4 Aug. 1902, 2:106; 25 Jan. 1912, 21:91, PR.

75. Entry, 7 Jan. 1900, 1:77; 23:48, PR.

76. Entry, 13 Dec. 1910, 18:3; entry, 8 Mar. 1915, 29:132; see similarly entry, 11 Oct. 1911, 20:104.

77. Entries, 20 Aug. 1897, 25 Feb. 1897, and 21 Aug. 1897, fld. "1896–1897," PR. Cabot's reflections on his own self-conscious anti-Semitism are examined further in Chapter 6. See also his comments in *Foregrounds and Backgrounds in Work with the Sick* (Boston: New England Hospital for Women and Children, 1906).

78. For contemporary observations on Boston's racial and ethnic categories, see Willard De Lue, "The Contribution of the Newer Races," in *Fifty Years of Boston*, ed. Herlihy, with a valuable interpretation of classification in Matthew Frye Jacobson, *Whiteness of a Different Color: European Immigrants and the Alchemy of Race* (Cambridge, Mass.: Harvard University Press, 1998). Racial categories in clinical notes of the period are handled deftly in Elizabeth Lunbeck, *The Psychiatric Persuasion: Knowledge, Gender and Power in Modern America* (Princeton, N.J.: Princeton University Press, 1994), 98–99, 122–23.

79. Cabot did see a couple of patients who were Irish, judging by their sur-names: entry, 8 Jan. 1906, 6:242, PR. Irish patients also figure occasionally into Cabot's published case reports, most prominently in illustrating the pathological effects of excessive alcohol use: Cabot, *Differential Diagnosis*, 320, 501.

80. Cabot, *Differential Diagnosis*, 344, 358.

81. Letter to RCC, 7 July 1910, 14:61; PR.

82. Letter to RCC, 6 Mar. 1911, 18:116 PR.

83. Sarah Bond, MD, letter to RCC, 22 Sept. 1905, 5:176, PR.

84. Dr Fullerton, letter to RCC, 8 Sept. 1906, 6:85, PR.

85. Letter to RCC, n.d., 23:48, PR.

86. Letter to RCC, 9 July 1915, 15:112; letter to RCC, 20 Sept. 1900, 1:256, PR.

87. Dr. W. C. MacDonald, letter to RCC, 30 Oct. 1906, 6:151, PR.

88. Christopher W. Crenner, "Professional Measurement: Quantification of Health and Disease in American Medical Practice, 1880–1920" (PhD diss., Harvard University, 1993), 33 n. 76.

89. Entries, 4 Mar. 1914, 27:32, 6 Mar. 1914, 27:33, 7 Mar. 1914, 27:34, 12 Mar. 1914, 27:47; entry, 3 Oct. 1912, 23:37; entry, 6 Jan. 1908, 8:153, PR.

90. Entry, 13 May 1913, 25:27, PR.

91. Entry, 6 Sept. 1906, 6:66 [bedwetting]; entry, 1 Sept. 1906, 6:50 [doctor]; entry, 24 Feb. 1914, 27:14 [wife doctor]; entry, 7 Oct. 1912, 23:43 [nurse]; entry, 9 Mar. 1914, 27:37 [relative], PR.

92. Konold, *History of American Medical Ethics*, 60.

93. Lee, *Happy Life of a Doctor*, 168. See an earlier discussion in James B. Herrick, *Memories of Eighty Years* (Chicago: University of Chicago Press, 1949), 156–60.

94. Hadra, *Public and the Doctor*, 133–36. See for another example T. J. Happel, "Quo Vadis?" *JAMA*, 1899, 32 (6): 271–75. See also Nathan E. Wood, *Dollars to Doctors or Diplomacy and Prosperity in Medical Practice* (Chicago: 1903). A secondary discussion appears in Konold, *History of American Medical Ethics*. See earlier reference in George Rosen, *Fees and Fee Bills: Some Economic Aspects of Medical Practice in Nineteenth Century America* (Baltimore: Johns Hopkins Press, 1946), 47, 58. Comparison to income tax appears in Hugh Cabot, *Doctor's Bill*, 122–23.

95. Konold, *History of American Medical Ethics*, 68–71.

96. Gerster, *Recollections of a New York Surgeon*, 254–57. Cf. Lee, *Happy Life of a Doctor*, 160–64. See also the secondary discussion in Starr, *Social Transformation of American Medicine*, 110–11.

97. Herrick, *Memories of Eighty Years*, 86–87. See another example in Lee, *Happy Life of a Doctor*, 168–70.

98. Cabot, "Better Doctoring," 7–9. Also see the continuation of the article in Richard C. Cabot, "Better Doctoring for Less Money: More Light on This Subject," *American Magazine*, 1916, 81 (5): 43–44, 76–78, 81.

CHAPTER 3. THE DIAGNOSIS OF HIDDEN DISEASE

1. Letter to RCC, 23 Feb. 1939, 34:70, PR.

2. Roger Lee, "Report," 29 May 1919, ibid.

3. General observations on the various uses of diagnosis in routine practice in this period appear in Barbara Sicherman, "The Uses of a Diagnosis: Doctors, Patients, and Neurasthenia," *Journal of the History of Medicine and Allied Sciences,* 1977, 32:33–54. A careful exploration of the complexity of twentieth-century diagnostic reasoning and the problems it presents in practice appeared in Ralph L. Engle, Jr., and B. J. Davis, "Medical Diagnosis: Present, Past, and Future," *Archives of Internal Medicine,* 1963, 112:512–43. The theoretical significance of the diagnosis as an organizing element of medical knowledge is outlined in Knud Faber, *Nosography: The Evolution of Clinical Medicine in Modern Times,* 2d ed. (New York: Paul B. Hoeber, 1930). Examples relevant to diagnosis in an earlier period appear in Malcolm Nicolson, "The Art of Diagnosis and the Five Senses," in *Companion Encyclopedia of the History of Medicine,* ed. William F. Bynum and Roy Porter (London: Routledge, 1993), 801–25.

4. James B. Herrick, "Modern Diagnosis: The Second Annual Alpha Omega Alpha Annual Lecture," *JAMA,* 1929, 92 (7): 518–22.

5. For general references on the history of medical technology in American medicine, see Chapter 1, n. 15. For the views of Cabot's day, cf. Henry L. Elsner, "On the Value to the Physician of Modern Methods of Diagnosis," *BMSJ,* 1902, 146:101–5. The general sentiment appears in Editorial, "Laboratory Investigation and Its Effects on Clinical Diagnosis," *JAMA,* 1900, 35:1679. On the need for practical experience in clinical diagnosis to balance textbook theories, see S. C. Gordon, "Common Sense in Medicine," *BMSJ,* 1891, 25:233–36, 261–63. For the need for scientific training in diagnosis for the practicing physician, see Victor C. Vaughan, "The Kind and Amount of Laboratory Work Which Shall Be Required in Our Medical Schools," *BMSJ,* 1892, 127:212–14.

6. "Obituary: Alfred Worcester," *JAMA,* 1951, 147:1292. Worcester characterized his own private practice in Waltham, Massachusetts, in a lengthy letter to Cabot when he was a medical student: A. Worcester to RCC, 23 Nov. 1895, fld. "Psychotherapy," box 71, Correspondence, Cabot Papers. Alfred Worcester, "Past and Present Methods in the Practice of Medicine," *BMSJ,* 1912, 166:159–64.

7. Worcester, "Past and Present."

8. Ibid., 163.

9. RCC, letter to F. Dewing, 10 May 1903, box 12, Correspondence, Cabot Papers.

10. Worcester, "Past and Present," 163, 159.

11. Francis W. Peabody, *Doctor and Patient* (New York: Macmillan, 1930) 23. This essay was reprinted from Francis Weld Peabody, "The Care of the Patient," *JAMA,* 1927, 88 (12): 877–82, which was in turn developed from an earlier lecture.

Details of Peabody's final illness are found eloquently set forth in Oglesby Paul, *The Caring Physician: The Life of Dr. Francis W. Peabody* (Boston: Francis A. Countway Library of Medicine et al., 1991), 121–41.

12. Peabody, *Doctor and Patient*, 36, 52.

13. Note (RCC autograph), "Important Events in the Field of Medicine, 1900–1910," n.d., box 76, General Files, Cabot Papers.

14. Richard Cabot, *Foregrounds and Backgrounds in Work for the Sick* (Boston: New England Hospital for Women and Children, 1906) unpaginated, [7]. The section on patients as diseases appears in a more extended discussion of social work in Richard C. Cabot, *Social Service and the Art of Healing* (New York: Moffat, Yard & Co., 1909) 33 ff.

15. The quotation is from Cabot, *Foregrounds and Backgrounds,* [4]. On Cabot as a progressive reformer, see Christopher Crenner, "Organizational Reform and Professional Dissent in the Careers of Richard Cabot and Ernest Amory Codman, 1900–1920," *Journal of the History of Medicine and Allied Sciences,* 2001, 56 (3): 211–37. For an excellent overview of Cabot's social reforms, see Ian S. Evison, "Pragmatism and Idealism in the Professions: The Case of Richard C. Cabot, 1869–1939" (PhD diss., University of Chicago, 1995). Robert Wiebe, *The Search for Order, 1877–1920* (New York: Oxford University Press, 1967), 111–32, emphasizes the affiliation of physicians with progressive reform as professional experts and managers in "the new middle class." But note that in the wake of more recent debate over the possibility of defining a single strand of political progressivism that James J. Connolly, *The Triumph of Ethnic Progressivism: Urban Political Culture in Boston, 1900–1925* (Cambridge, Mass.: Harvard University Press, 1998), examining very similar materials, finds evidence of a reversal in what he sees as the assumed relationship between political and ethnic identity in the standard analysis and extends the notion of progressivism to a wider array of groups and activities.

16. On the physician's personal service, see Cabot, *Foregrounds and Backgrounds,* [4]. Remarks in a similar vein appear in Richard Cabot, "Suggestions for the Reorganization of Hospital Out-Patient Departments with Special Reference to the Improvement of Treatment," *Maryland Medical Journal,* 1907, 50 (3): 81–91. On the establishment of medical social work, see Richard Cabot, *Social Work: Essays on the Meeting Ground of Doctor and Social Worker* (Boston: Houghton, 1919) vii–xxvii. Also see Richard Cabot, "Social Work: The Diagnosis and Treatment of Character in Difficulty," *Charities and the Commons,* 1902, 19:1001–10. On the responsibility for a renewed therapeutics, see Richard Cabot, "The Renaissance of Therapeutics," *JAMA,* 1906, 46 (22): 1660–65.

17. Cabot's founding role in the Clinicopathological Conferences carried enough symbolic weight to bear celebrating on the occasion of the first new case of the millennium: Robert E. Scully, "Case 1-2000," *NEJM,* 2000, 34 (2): 115. Cabot's recollections of the origin of the conference appear in Frederic Washburn, *The Massachusetts General Hospital: Its Development, 1900–1935* (Boston: Houghton Mifflin, 1939), 115–16. An account of an earlier version of the conferences in the form

of a teaching exercise appears in Richard Cabot, "A Case-History Clinic," *Detroit Medical Journal,* 1905, 5:6–13. Later accounts of the conferences noting the addition of the use of hidden autopsy information are to be found in Richard C. Cabot and Russell Dicks, *The Art of Ministering to the Sick* (New York: Macmillan, 1936), 323–24. See also the mention in Richard Cabot, "A Study of Mistaken Diagnoses Based on an Analysis of 1,000 Autopsies and a Comparison with the Clinical Findings," *JAMA,* 1910, 55 (16): 1343–50. On the circulation of reports before formal publication, see RCC to Frederic Washburn, 19 Mar. 1915, fld. "Letters from Cabot, 1908–1929," box 16, Correspondence, Cabot Papers. The first published version of the conferences is found in Richard C. Cabot and Hugh Cabot, "Antemortem and Postmortem Records as Used in the Weekly Clinicopathological Exercises, 10271," *BMSJ,* 1924, 191 (4): 30–39.

18. "Report of Clinical Conferences under Professor Shattuck," vol. 2 (3 Oct. 1890-29 Apr. 1892), p. 105, Manuscript Collection, Countway. Cabot mentions Shattuck's course in one account of the origins of the Clinicopathological Conference (Washburn, *Massachusetts General Hospital,* 115–16), but not his own role as a successful participant.

19. Cabot, "Antemortem and Postmortem," 39.

20. On differential diagnosis, see Richard C. Cabot, *Case Teaching in Medicine: A Series of Graduated Exercises in Differential Diagnosis, Prognosis and Treatment of Actual Cases of Disease* (Boston: D. C. Heath, 1906). For an independent contemporary discussion of differential diagnosis, see Glenworth Butler, *The Diagnostics of Internal Medicine: A Clinical Treatise upon the Recognized Principles of Medical Diagnosis* (New York: Appleton, 1901). The endorsement of the importance of differential diagnosis in American textbooks dated to the mid nineteenth century: see, e.g., Austin Flint, *A Practical Treatise on the Physical Exploration of the Chest, and the Diagnosis of Diseases Affecting the Respiratory Organs,* 2d ed. (Philadelphia: H. C. Lea, 1866), 101–2.

21. Richard C. Cabot, "A Study of Mistaken Diagnoses: Based on the Analyses of 1000 Autopsies and a Comparison with the Clinical Findings," *JAMA,* 1910, 55 (October): 1343–50.

22. Remarks of Ida M. Cannon speaking at Memorial Seminar, transcript, 18 Jan. 1958, fld. "Cabot Seminar Papers," box "Memorial Seminar," Cabot Papers.

23. The bottom line of each record for a new patient's visit ends with an entry of the diagnosis. The last pages of each volume of records contain an index to the diagnoses on all new patients in that volume. There were almost no discrepancies between the two lists, but they were sometimes modified later to correct differences.

24. Cabot's private charts look similar in format to the patient records of the hospital where he began his training. See, e.g., records from the period in MGH, PR, selected volumes 1901 to 1920. Other examples are cited from similar records at the Hospital of the University of Pennsylvania and Johns Hopkins Hospital in Christopher W. Crenner, "Professional Measurement: Quantification of Health

and Disease in American Medical Practice, 1880–1920" (PhD diss., Harvard University, 1993).

25. Letter to RCC, June 1908, 2:58, PR.

26. Letter to RCC, June 1908, 1:149, PR. Letter to RCC, June 1908, 2:58, PR. The similar dates of the two letters resulted from a general mailing Cabot made to his patients in 1908 requesting information on "the outcome of the illness for which I saw you"; example of standard letter: 2:58, PR.

27. Letter to RCC, 24 May 1913, 25:40, PR.

28. Ibid. On Cabot's controversial role in debates over the disclosure of life-threatening diagnoses to patients, see Stanley Joel Reiser, "Words as Scalpels: Transmitting Evidence in the Clinical Dialogue," *AIM*, 1980, 92 (6): 837–42. See also Chester R. Burns, "Richard Clarke Cabot (1868–1939) and Reformation in American Medical Ethics," *BHM*, 1977, 51:353–68.

29. See n. 24 above.

30. Entry, 5 Sept. 1903, 2:268, PR. For a similar description of angina from the period, see William Osler, *Principles and Practice of Medicine: Designed for the Use of Practitioners and Students of Medicine (1892), 4th ed.* (New York: D. Appleton, 1902), 761–65.

31. Entry, 5 Sept. 1903, 2:268; entries, 15 May 1900, 27 May 1900, and 1 Aug. 1900, 1:104, PR.

32. Letter to RCC, June 1908, 2:79; letter to RCC, 7 May 1902, 1:168; letter to RCC, June 1908, 2:209; letter to RCC, 20 Jan. 1903, 2:163, PR.

33. Letter to RCC, 31 Dec. 1914, 1:65, PR.

34. Letter to RCC, 23 Mar. 1909, 7:23, PR.

35. Cabot recalled that he got his first patients from other doctors for whom he did blood testing: "Autobiographical Notes," p. 8, fld. "Autobiographical Notes," box 109, Biographical Material, Cabot Papers. See Richard C. Cabot, *Physical Diagnosis of the Chest* (New York: W. Wood & Co., 1900). Also see Richard C. Cabot, *Exercises in Differential Diagnosis* (Boston: Groom, 1902). The widely circulated second edition is Richard C. Cabot, *Differential Diagnosis* (Philadelphia: W. B. Saunders, 1912). For the use of referrals, see Chapter 2 above.

36. See, e.g., a case where other doctors "could not make a very definite diagnosis": Dr. P. A. Adamian, letter to RCC, 17 Feb. 1913, 24: 61, PR.

37. Letter to RCC, 15 Dec. 1903, 2:249; letter to RCC 23 June 1909, 11:5; entry, 23 Apr. 1912, 18:20; entry [August 1899], 1:47; letter to RCC, June 1908, 2:158; entry September 1900, 1:150, PR.

38. Entry, January 1902, 1:219; and entry September 1900, 1:150, PR. The entries for this woman took up enough room in the chart to extend over several pages.

39. Letter to RCC, 11 Feb. 1913; also Dr. [F?]owle, letter to RCC, 1 Apr. 1912, 18:99, PR.

40. Letter to RCC, June 1908, 2:58; letter to RCC, June 1908, 1:149; letter to RCC, n.d., 1:114; letter to RCC, June 1908, 2:261; letter to RCC, 25 Jan. 1912, 18:82;

letter to RCC, 13 June 1908, 3:74; letter to RCC, June 1908, 3:74; letter to RCC, 28 June 1913, 18:92; letter to RCC, June 1908, 2:58; letter to RCC, June 1908, 2:261; letter to RCC, n.d., 1:114, PR. Letter to RCC, 10 Aug. 1908, fld. "Miss Mavis Hunt;" box 71, Correspondence; letter Dr. Albert Tenney to RCC, 9 Aug. 1909, fld. "T," box 35, Correspondence, Cabot Papers. For an example of slang for illness, see Frederic G. Cassidy, ed., *Dictionary of American Regional English,* vol. 2 (Cambridge, Mass.: Belknap Press, 1991), 100.

41. Letter to RCC, n.d., 9:121, PR.

42. Letter to RCC, June 1908, 2:58; letter to RCC, June 1908, 1:149; letter to RCC, n.d., 1:114; letter to RCC, June 1908, 2:261; letter to RCC, 25 Jan. 1912, 18:82; letter to RCC, 13 June 1908, 3:74; letter to RCC, June 1908, 3:74; letter to RCC, 28 June 1913, 18:92; letter to RCC, June 1908, 2:58; letter to RCC, June 1908, 2:261; letter to RCC, n.d., 1:114, PR.

43. Letter to RCC, 10 Mar. 1907, 6:37, PR.

44. This distinction was not one named by these patients but draws on a valuable analysis well articulated in Owsei Temkin, "The Scientific Approach to Disease: Specific Entity and Individual Sickness," in id., *The Double Face of Janus and Other Essays in the History of Medicine* (Baltimore: Johns Hopkins University Press, 1977), 441–55.

45. Letter to RCC, 21 Oct. 1913, 26:37; letter to RCC, 27 Oct. 1916, 32:96; letter to RCC, 21 Oct. 1913, 26:37; letter to RCC, 1 Feb. 1914, 25:40, PR.

46. Richard C. Cabot, *A Layman's Handbook of Medicine: With Special Reference to Social Workers* (Boston: Houghton Mifflin, 1916), 77–78.

47. George Bernard Shaw, *The Doctor's Dilemma; a Tragedy* (Baltimore: Penguin Books, 1913), 7. The first American publication was George Bernard Shaw, *The Doctor's Dilemma, Getting Married, and the Shewing up of Blanco Posnet* (New York: Bretano's, 1911). See details of his position in Richard C. Cabot, "Doctor's Dilemma in Bernard Shaw and in Fact," *Survey,* 1911, 26 (6): 381–83. Other physicians were not pleased at Cabot's support of Shaw, as evidenced in Editorial, "Cabot, Bernard Shaw and the Medical Profession," *JAMA,* 1915, 164 (9): 762–63.

48. For the pleasure of being not needed, see Cabot, *Social Work,* 98. For the support of the public, see, e.g., the unsigned letter to RCC, n.d., Falls River, Massachusetts, fld. "A," box 35, Correspondence, Cabot Papers. For the creation of disease, see entry, 6 Jan. 1909, 11:105, PR.

49. Tabulation of all new-patient diagnoses by Cabot in 1915 showed approximately 7 percent with the notation "not ill." The rate was about 6 percent of new cases for 1912–14. For more detail, see Chapter 5. See, e.g., records for patients diagnosed as not ill in 16:71, 77, 89, 92, 101, 123, 149, PR. Cabot generally included future visits with the same patient on the same page with the original visit. A notable exception was a man who had two completely negative evaluations separated by fifteen years. See entries, 24 Oct. 1899, and 7 Oct. 1914, 1:65, PR

50. An exception can be found, e.g., in the unusual case of a 21-year-old woman who felt perfectly well but who was sent to Cabot by her doctor under the

pretext of getting a routine blood exam. Her doctor suspected a serious heart condition, which this doctor wanted Cabot to confirm for her without disclosing this intent to the patient. See entry, 24 June 1912; and letter to RCC, 23 June 1912, 23:12, PR.

51. Elwood Worcester, Samuel McComb, and Isador H. Coriat, *Religion and Medicine: The Moral Control of Nervous Disorders* (New York: Moffat, Yard, & Co., 1908), 5.

52. B. A. Souders, "Letter to the Editor," *Medical World,* 1911, 29 (3): 102–3. See also Editorial, "Physicians and the Emmanuel Movement," *Northwest Medicine,* 1909, 7:20–21.

53. Letter to RCC, 1 Aug. 1899, 1:31; letter to RCC, 14 Jan. 1912, 15:18; letter to RCC, 11 Feb. 1913, 18:99, PR.

54. "Report," MGH, October 1914, 1:65; and entry, 1:65, PR. A related account of a patient's case appears in Patricia Spain Ward, "The Medical Brothers Cabot: Of Truth and Consequences," *Harvard Medical Alumni Review,* 1982, 56 (4): 30–34.

55. Letter to RCC, 31 Dec. 1914, 1:65, PR.

56. For an excellent preliminary review of the question, see Allan M. Brandt and Martha Gardner, "The Golden Age of Medicine?" in *Medicine in the Twentieth Century,* ed. Roger Cooter and John Pitchstone (Amsterdam: Harwood Academic Publishers, 2000), 21–37.

CHAPTER 4. TREATMENT

1. Entry, 24 Feb. 1911; letter to RCC, 11 Feb. 1913; and letter to RCC, 1 Apr. 1912, 18:99, PR.

2. Letter to RCC, June 1908, 2:205, PR.

3. RCC, letter to Dr. G. S. Foster, 4 Oct. 1912, 18:6, PR.

4. See an example of coping with this difficulty in S. L. Abbot, "Hepatic Abscess; Fatal Peritonitis," *BMSJ,* 1875, 92 (15): 429–37.

5. Memorial Seminar, transcript, 18 Jan. 1958, fld. "Cabot Seminar Papers," box "Memorial Seminar," Cabot Papers.

6. Sarah Bond, MD, letter to RCC, 7 Dec. 1915, 31:56, PR.

7. E. Libman, letter to RCC, 10 Mar. 1903, 2:175, PR.

8. N. S. Davis, "Need of Much More Accurate Knowledge Concerning Both the Immediate and Remote Effects of the Remedial Agents in General Use," *JAMA,* 1902, 38 (22): 1415–18. Cf., e.g., Abbot, "Hepatic Abscess." The latter article gave a wonderfully detailed account of an extended course of medications and treatments in the nineteenth century aiming successively to improve sleep, stimulate appetite, expel gas, soothe pains, quell fevers, and induce bowel movements, each judged in turn by these intended effects as confirmed by the patient.

9. On the general theories underlying nineteenth-century therapeutics, see John Harley Warner, *The Therapeutic Perspective: Medical Practice, Knowledge, and Identity in America, 1820–1885* (Cambridge, Mass.: Harvard University Press, 1986).

See also Charles E. Rosenberg, "The Therapeutic Revolution: Medicine, Meaning, and Social Change in 19th-Century America," in *Explaining Epidemics and Other Studies in the History of Medicine,* ed. Charles E. Rosenberg (New York: Cambridge University Press, 1997), 9–31, and the essays in Morris J. Vogel and Charles E. Rosenberg, eds., *The Therapeutic Revolution: Essays in the Social History of American Medicine* (Philadelphia: University of Pennsylvania Press, 1979).

10. Charles M. Whitney, "The Impossibility of Curing Syphilis by Salvarsan Alone and the Dangers Arising from Insufficient Treatment," *Interstate Medical Journal,* 1916, 23:80–88.

11. Ibid., 88.

12. Ibid. See also Editorial, "The Use and Abuse of Salvarsan," *Journal of the Minnesota State Medical Society,* 1911, 31 (21): 524–25.

13. George M. Dock, "The Movement for Exact Treatment," *Journal of the Iowa State Medical Society,* 1912, 2 (2): 89–94. Echoing this sentiment about the need to enlist a patient to follow therapeutic advice is G. D. Lind, "Popular Delusions as Affecting the Physician," *West Virginia Medical Journal,* 1912, 6 (8): 261–63.

14. Dock, "Movement for Exact Treatment," 94.

15. Ibid.

16. Whitney, "Impossibility," 87.

17. Ibid.

18. Albert C. Geyser, "Physiological *Versus* Symptomatic Therapy," *New York Medical Journal,* 1916, 103:18–21.

19. Whitney, "Impossibility," 87.

20. Oliver Wendell Holmes, "The Young Practitioner," in id., *Medical Essays, 1842–1882* (Boston: Houghton, Mifflin, 1888), 370–95. This graduation address from 1871 succeeded in playing more gently on this theme than usual. For the theoretical compensations of "behind-the-scenes" ridicule by agents who feel they must accommodate their audience, see Erving Goffman, *The Presentation of Self in Everyday Life* (Garden City: Doubleday Anchor, 1959), 170–75, who cites the case of physicians. To some extent, doctors in vastly different periods and cultures have shared similar problems in enforcing therapeutic recommendations. The Hippocratic corpus belittles the reliability of patients and warns doctors against trusting that they will take treatments: Ludwig Edelstein, "The Hippocratic Physician," in *Ancient Medicine: Selected Papers of Ludwig Edelstein,* ed. O. Temkin and C. L. Temkin (Baltimore: Johns Hopkins University Press, 1931), 87–110.

21. Letter to RCC, 19 Feb. 1915, 22:16; letter to RCC, June 1908, 2:209; letter to RCC, 28 June 1909, 9:86; letter to RCC, 29 Nov. 1909, 14:100; to RCC, June 1908, 2:205, PR.

22. Entry, 13 Mar. 1908, 9:137; letter to RCC, 25 Jan. 1912, 18:92; entry, 26 Oct. 1916, 32:9; letter to RCC, 7 May 1902, 1:168; and letter to RCC, n.d., 15:47, PR.

23. Entry, 20 June 1916, 6:16; letter to RCC, 10 June 1908, 6:8; letter to RCC, 11 June, 1908, 6:37, PR.

24. Letter to RCC, 11 June 1908, 6:37, PR.

25. Harry M. Marks, *The Progress of Experiment: Science and Therapeutic Reform in the United States, 1900–1990* (New York: Cambridge University Press, 1997), 17–41.

26. Geyser, "Physiological *Versus* Symptomatic Therapy," 21.

27. Edward E. Cornwall, "Some Aspects of Symptomatic Treatment," *New York Medical Journal*, 1918, 107:974–76, quotation from 974.

28. J. B. McGee, "The Present Status of Therapeutics," *Cleveland Medical Journal*, 1912, 11 (7): 496–500.

29. Ibid., 498.

30. Editorial, "Extremes in Theory and Practice," *American Medicine*, 1903, 5 (2): 41.

31. Richard C. Cabot, "Therapeutics Based on Pathological Physiology," *BMSJ*, 1906, 155 (3): 57–59. For similar ideas about the rationale of treating symptoms, cf. A. Barnes Hooe, "Rational Versus Empirical Therapeutics," *Washington Medical Annals*, 1905, 4:200–205; Edward E, Cornwall, "Some Aspects of Symptomatic Treatment," *New York Medical Journal*, 1918, 107:974–76. Or in a slightly earlier version, see J. C. Applegate, "Treatment of Diseases Versus Treatment of Symptoms," *Medical Bulletin*, 1892, 14:297–300.

32. Richard C. Cabot, "Therapeutics Based on Pathological Physiology," *Transactions of the Association of American Physicians*, 1906, 21:187–92.

33. Wallace C. Abbott, "A Plea for a Truer Therapy-Real Treatment of the Sick," *New York Medical Journal*, 1904, 29:1238–40. The same metaphor appears, e.g., in Torald Sollmann's famous *Manual of Pharmacology*, 2d ed. (Philadelphia: W. B. Saunders, 1922), 71–72; and Hobart Amory Hare, *A Text-Book of Practical Therapeutics*, 7th ed. (Philadelphia: Lea Brothers, 1898), 19.

34. Hippocrates, *Epidemics* 1.11.14–15, ed. W. H. S. Jones (Cambridge, Mass.: Harvard University Press, 1923), 164. The translation is mine.

35. Nicholas A. Christakis, *Death Foretold: Prophecy and Prognosis in Medical Care* (Chicago: University of Chicago Press, 1999), 4–8, 130–34. Christakis finds a long-standing "complementarity of prognosis and therapy" in which improvements in medical therapy over the twentieth century led to a declining emphasis on prognosis. I find this argument compelling, namely, that the historical development of powerful, specific therapies shifted attention away from prognosis. As the Hippocratic "Prognosis" sensibly pointed out two millennia earlier, it was always better to cure people than to foresee the course of their sickness. I am expanding here on Christakis's observation, noting that prognosis, before it yielded ground to therapeutics, functioned to supplement and support it.

36. Dr. H. M. Chapman, letter to RCC, 25 June 1912, 23:17, PR.

37. G. S. Foster, MD, letter to RCC, 26 Mar. 1914, 23:35, PR.

38. Letter to RCC, 9 Jan. 1912, 18:62, PR, emphasis in the original. Dr. William A. Bell, letter to Dr. William W. Gannett, 12 September 1886, tipped into practice notebooks, "Volume I," p. 255, Gannett Papers: "He is kept in certain measure

ignorant of his real condition—against my advice but by the wishes of his family," in regard to a patient with an "unfavorable prognosis."

39. Daniel W. Cathell, *The Physician Himself and Things That Concern His Reputation and Success*, 10th ed. (Philadelphia: F. A. Davis, 1900), 173.

40. [Physician], letter to [patient's brother], 6 Mar. 1915, 16:42, PR.

41. Letter to RCC, 1 Feb. 1914, 25:40, PR. Cf. the letter from a woman with severe tuberculosis who wrote to Cabot, "It is very difficult when a patient does not understand technical terms and until today I never would get Dr. Penny to talk it frankly out with me." Letter to RCC, n.d., 9:121, PR.

42. Letter to RCC, 31 Dec. 1914, 1:65, PR.

43. Richard C. Cabot, "The Use of Truth and Falsehood in Medicine: An Experimental Study," *American Medicine*, 1903, 5 (9): 344–49. In another context, see Richard C. Cabot, "The Doctor and the Community," *American Medicine*, 1904, 8 (17): 731–32. Cf. the reference to medical practices in Richard C. Cabot, "Justifiable Lying," *Journal of Education*, 1909, 69 (6): 145–46. Cabot would directly subvert another doctor's efforts to manipulate a prognosis in order to gain cooperation with a plan of treatment: letter to RCC, 15 Dec. 1910; letter to RCC, 2 Dec. 1913; and entry, 16 Dec. 1910, 18:6, PR. Letter to RCC, 10 Feb. 1915; and entry, 1 May 1912, 22:84, PR. Although Cabot noted the diagnosis of "gastric cancer" for this patient, the daughter wrote him later that the patient remained "ignorant to the end of the true cause of her illness," believing in part that her pains were from "rheumatism." In addition, the daughter reminded Cabot that her mother had been for a time "very hopeful after visiting your office." In addition, Cabot gave moving testimony, however, about his efforts to conceal from his wife his discovery of her fatal breast cancer, as will be considered further in a subsequent chapter.

44. Letter to RCC, 21 June 1914; RCC, letter to [patient], n.d., 27:48, PR.

45. RCC, letter, 21 Oct. 1912, 23:53, PR.

46. See, e.g., Howard Brody, *Placebos and the Philosophy of Medicine: Clinical, Conceptual, and Ethical Issues* (Chicago: University of Chicago Press, 1980). See Stanley Joel Reiser, "Words as Scalpels: Transmitting Evidence in the Clinical Dialogue," *AIM*, 1980, 92 (6): 837–42. See also Howard M. Spiro, *Doctors, Patients, and Placebos* (New Haven, Conn.: Yale University, 1986), 121–23.

47. Remarks of Paul Dudley White at Memorial Seminar, transcript, 18 Jan. 1958, fld. "Cabot Seminar Papers," box "Memorial Seminar," Cabot Papers.

48. Remarks of Dr. Young, "Memorial Seminar," 21 ff. Cabot's office records documented a case of a prominent woman whom Cabot diagnosed with cervical cancer on the basis of interview and physical examination: entry, December 1914, 29:44, PR. Included with the case was the later report of a biopsy done by Hugh Cabot, January 1915, demonstrating a benign cervical growth.

49. Letter to RCC, June 1908, 2:97, PR.

50. Letter to RCC, 31 Mar. 1910, 15: 99, PR.

51. Letter to RCC, 1 June 1908, 1:174, PR.

52. Letter to RCC, 31 Jan. 1904, 2:265, PR.

53. Letter to RCC, 19 Oct. 1900, 1:143, PR. See similarly a woman who speculates whether her symptoms are from her disease or "due to the large amount of iron" she had taken to treat it: letter to RCC, 24 Jan. 1911, 18:59, PR.

54. Letter to RCC, 8 Apr. 1912, 22:9, PR.

55. Letter to RCC, 19 May 1902, 2:77, PR.

56. Letter to RCC, 16 Oct. 1912, 23:40, PR.

57. Agnes C. Victor, MD, letter to RCC, 9 Oct. 1912, 23:40, PR.

58. Entries, 19 and 23 Dec. 1898, 1:13, PR.

59. Dr. JSC, letter to RCC, 27 Mar. 1919, 34:52, PR.

60. Entries, 21 and 29 Oct. 1899, 1:13, PR. Cabot used the medicine occasionally after this incident but noted in a similar case later that "Urotropine [sic] did harm"; he seems to have dropped its use subsequently: entry, 1 May, 1902, 2:79, PR.

61. S. J. Meltzer, "The Present Status of Therapeutics and the Significance of Salvarsan," *JAMA*, 1911, 56 (23): 1709–13.

62. Ibid., 1709.

63. Fielding H. Garrison, "Ehrlich's Specific Therapeutics in Relation to Scientific Method," *Popular Science Monthly*, 1911, 78 (15): 209–22.

64. Meltzer, "Present Status of Therapeutics," 1709.

65. Entry, 1 May 1902; letter to RCC, 19 May 1902; [patient], note to RCC, "Tuesday morn."; letter to RCC, "Tuesday evening," 2:77, PR.

66. Note "Tuesday morn.," 2:77, PR.

67. Entry, 17 May 1902, 2:77, PR.

68. Entries, 1, 17, and 22 May 1902; and letter to RCC, 19 May 1902, 2:77, PR. Emphasis in original.

69. William V. McDermott, *Surgery at the New England Deaconess Hospital, 1896–1985* (Boston: Countway Library of Medicine, 1995), 5–6; he quotes a local religious magazine's praise of the hospital's operating room, which boasted "plate glass shelves and three fine Welsbach lights and an operating table of the latest design and finish" (6).

70. Edward Reynolds, "Nephroureterectomy for Tuberculous Disease with a Description of a New Technic," *American Medicine*, 1905, 9:313–17.

71. Reynolds, letter to RCC, 26 Oct. 1902, 2:79, PR.

72. Reynolds, letter to RCC, 17 Sept. 1903, 2:79, PR.

73. Reynolds, "Nephroureterectomy," 313.

74. Letter to RCC, June 1908, 2:79, PR.

75. Ibid.

76. Letter to RCC, 19 May 1902, 2:79, PR.

77. Letter to RCC, June 1908, 2:54, PR.

78. Letter to RCC, June 1908, 2:205, PR.

79. Letter to RCC, June 1908, 2:209; and letter to RCC, 30 Jan. 1904, 2:222, PR.

80. Letter to RCC, 4 Apr. 1902, 2:55, PR.

81. Letter to RCC, 25 Mar. 1902, 2:55, PR.

82. Letter to RCC, 22 June 1915, 15:112, PR.

83. Letter to RCC, 6 Mar. [1915], 22:144, PR.

84. Letter to RCC, June 1908, 2:205, PR.

85. Richard C. Cabot, *Differential Diagnosis,* 2d ed. (Philadelphia: W. B. Saunders, 1912), 540–42, 570–71. See a similar assessment of pernicious anemia the same year in William Osler, *The Principles and Practice of Medicine, 8th ed.* (New York: D. Appleton, 1912), 733–39.

86. Letter to RCC, 31 Oct. 1903, 2:235, PR. The number that Cabot entered in the chart was 1,448,000, which his patient seems to have rounded off: entry, 20 July 1903, 2:235, PR.

87. Letter to RCC, 1 Feb. 1904, 2:235, PR.

88. Ibid.; entry, 20 July 1903, 2:235; and letter to RCC, 28 Nov. 1904, with entry, 19 Apr. 1904, 3:115, PR.

89. Entry, 21 Oct. 1912; letter to RCC, 13 Nov. 1915; Dr. Dudley Fulton, letter to RCC, 29 Jan. 1914; reports from "Physician's Clinical Laboratory," 7 May and 7 July 1913; all in 23:62, PR.

90. Cabot, "Therapeutics Based on Pathological Physiology," 187.

91. Arthur K. Shapiro and Elaine Shapiro, *The Powerful Placebo: From Ancient Priest to Modern Physician* (Baltimore: Johns Hopkins University Press, 1997), 175. A similar version appears in Spiro, *Doctors, Patients, and Placebos,* 121–23. Also see Brody, *Placebos and the Philosophy of Medicine.*

92. Cabot, "Uses of Truth and Falsehood."

93. Henry K. Beecher, "The Powerful Placebo," *JAMA,* 1955, 159 (17): 1602–6. Beecher's articles provided a coherent statement of the notion of the placebo as a research tool with important measurable effects. See a fine example of the early implications of these trials in Henry K. Beecher, "Surgery as Placebo: A Quantitative Study of Bias," *JAMA,* 1961, 176 (13): 1102–7. On recent controversial attempts to distinguish experiment more sharply from treatment, cf., e.g., Matthew Miller, "Phase I Cancer Trials: A Collusion of Misunderstanding," *Hastings Center Report,* July–August 2000, 30 (4): 34–43; and Robert J. Wells "Letter to the Editor: Phase I Cancer Trials: Therapeutic Research?" ibid. January–February 2001, 4.

94. Cabot, "Uses of Truth and Falsehood," 348. See James Peter Warbasse, *Medical Sociology* (New York: D. Appleton, 1909), 236.

95. Cabot, "Renaissance in Therapeutics," 1662.

96. Partly this is a characteristic of the medical literature, and partly a feature of the changing use of the term "placebo." It is striking, however, that while placebo treatments are discussed in the medical literature, I have been unable to find the term "placebo" in the *Index Medicus* or the *JAMA* subject index in the decades immediately before and after the turn of the century.

97. Robley Dunglison, *Medical Lexicon: A Dictionary of Medical Science* (Philadelphia: Lea & Blanchard, 1846), 587. See John Redman Coxe, *The Philadel-*

phia Medical Dictionary (Philadelphia: Thomas Dobson, 1808). See also David B. Morris, "Placebo, Pain, and Belief: A Biocultural Model," in *The Placebo Effect*, ed. Anne Harrington (Cambridge, Mass.: Harvard University Press, 1997), 187–207.

98. J. Ware, "Dr. Ware's Lecture on General Therapeutics," *BMSJ*, 1862, 64:367–372, quotation from 372. And see George Rowland, "Practical Notes from a General Practitioner of Medicine," *Journal of the Indiana State Medical Association*, 1908, 1 (3): 95–97, which discusses placebos, if indirectly, on 96. See also Cathell, *Physician Himself* (1900), 150–51.

99. See Jurich v. General Motors Corp., 539 S. W. 2d 595 (Mo. Ct. App. 1976). This case includes testimony about the use of placebos by doctors and includes expert testimony on the nondisclosed use of medical placebos. I am indebted to Jerry Menikoff for bringing this case to my attention.

100. See, e.g., entry for 12 Jan. 1901, 1:194, PR: "Rx: placebo." Cabot seems to have used placebos very similarly later in his practice too, after the denunciation, although without the self-conscious identification of their use. See entry, 24 June 1912, 23:13, PR.

101. Anne Digby, *Making a Medical Living: Doctors and Patients in the English Market for Medicine, 1720–1911* (New York: Cambridge University Press, 1994), 199–212. See Dorothy Porter and Roy Porter, *Patient's Progress: Doctors and Doctoring in Eighteenth-Century England* (Oxford: Polity Press, 1989), 161–67. See also Shapiro and Shapiro, *Powerful Placebo*.

102. Francis Weld Peabody, "The Care of the Patient," *JAMA*, 1927, 88 (12): 877–82.

103. W[illiam] R[ichardson] Houston, "The Doctor Himself as a Therapeutic Agent," *AIM*, 1938, 11:1416–25. For biographical background, see AMA, *American Medical Directory*, 17th ed. (Chicago: AMA, 1942), 1788. A lengthier version of the argument, and somewhat diluted, appeared earlier in William R. Houston, *The Art of Treatment* (New York: Macmillan, 1936).

104. Cabot, "Uses of Truth and Falsehood," 348.

105. Rowland, "Practical Notes," 97.

106. E. A. Cockayne, George E. S. Ward, letter to the editors, *Lancet*, 1920, 1:221.

107. F. Ready, "The Justifiability of Therapeutic Lying," *Lancet*, 1920, 1:289. The editors at the *Lancet* initially treated this controversy as a balanced debate, although their sympathies seemed to shift to favor Carlill's opposition. The first set of letters responding to Carlill were indexed in the 1919 volume under "therapeutic" but the second set in the index in 1920 appeared under "lying."

108. Ready, "Justifiability," 289.

109. Ibid.

110. The application of detailed psychodynamic analysis to the ordinary medical relationship came to be epitomized in the work of Michael Balint in Britain in the 1950s and subsequently. Michael Balint, *The Doctor, His Patient, and the Illness* (New York: International Universities Press, 1957).

111. Houston, "Doctor Himself," 1422.

112. Ibid., 1425. See the favorable influence of Peabody in Houston, *Art of Treatment*, 75. Houston makes no mention of Cabot, although their interests overlapped very closely.

113. Spiro, *Doctors, Patients, and Placebos*, 121–23. Spiro gives a compelling modern version of a similar approach to the placebo and offers an ad hominem defense against Cabot's criticism.

114. Houston, "Doctor Himself," 1425. See Francis W. Peabody, *Doctor and Patient* (New York: Macmillan, 1930), 44–46.

115. Letter to RCC, 4 Nov. 1889, and see the series of related correspondence between November 1889 and December 1891, fld. "G," Personal Correspondence, Cabot Papers.

116. Letter to RCC, 31 May [1903?], ibid.

CHAPTER 5. NERVOUS DISEASE AND PERSONAL IDENTITY

1. Entry, 10 June 1902, 2: 88; and 5: 209; 8: 65; 12: 55; 15: 19, PR, Cabot Papers.

2. William Osler, *The Principles and Practice of Medicine* (New York: D. Appleton, 1892), 967–85. Osler maps these overlapping states of functional nervous disorders in contiguous sections on hysteria, neurasthenia, and neurosis.

3. Ibid., 978. A thorough discussion of the changing nature of medical theory on nervous hysteria appears in Mark S. Micale, *Approaching Hysteria: Disease and Its Interpretations* (Princeton, N.J.: Princeton University Press, 1995). For specific ideas about the relationship of these changes to practice, see Marijke Gijswijt, "Introduction: Cultures of Neurasthenia from Beard to the First World War," in *Cultures of Neurasthenia from Beard to the First World War*, ed. Marijke Gijswijt-Hofstra and Roy Porter (New York: Rodopi, 2001), 1–30.

4. Pierre Janet, *Les Obsessions et la psychasthénie*, 2 vols. (Paris: Felix Alcan, 1903). See also id., *The Major Symptoms of Hysteria: Fifteen Lectures Given in the Medical School of Harvard University* (New York: Macmillan, 1907). Janet was playful about the nomenclature, telling his American colleagues, "[Y]ou will call [these patients] neurasthenic for the family if you like" (ibid., 11).

5. Cabot announced his interest in Freudian psychoanalysis as early as 1906 in an address to the Maryland Medical Society, and he was alert to Freud's reception among his American colleagues: "Suggestions for the Reorganization of Hospital Out-Patient Departments with Special Reference to the Improvement of Treatment," *Maryland Medical Journal*, 1907, 50 (3): 81–91. See the full analysis of this topic in Nathan G. Hale Jr., *Freud and the Americans: The Beginnings of Psychoanalysis in the United States, 1876–1917* (New York: Oxford University Press, 1971). See the discussion also in F. G. Gosling, *Before Freud: Neurasthenia and the American Medical Community, 1870–1910* (Urbana: University of Illinois Press, 1987).

6. Charles Dana, "The Partial Passing of Neurasthenia," *BMSJ*, 1904, 150:

339–44. There is an extensive secondary literature on Boston particularly. Especially see Hale, *Freud and the Americans*, 88.

7. The use of the diagnosis in routine practice is considered in Barbara Sicherman, "The Uses of a Diagnosis: Doctors, Patients, and Neurasthenia," *Journal of the History of Medicine and Allied Sciences*, 1977, 32:33–54. A comparative look at changing notions of neurasthenia appears in Gijswijt, "Introduction: Cultures of Neurasthenia."

8. Entries for 14 Jan. 1910 and 10 Nov. 1911, 15:19, PR.

9. Janet Oppenheim, *"Shattered Nerves": Doctors, Patients, and Depression in Victorian England* (New York: Oxford University Press, 1991).

10. Cabot was one of four Boston physicians who helped in the founding of the Emmanuel Movement. For a patient referred by Cabot from his office to the group in 1908, see 11:22, PR. The movement collapsed as an organized enterprise in 1909, but during its last days, Cabot continued to offer qualified support in, e.g., "Suggestion, Authority, and Command," *Psychotherapy*, 1909, 2 (3): 17–29, and "The Literature of Psychotherapy," ibid., 3 (4): 18–27. See for the general outline of the movement: Hale, *Freud and the Americans*, 228–49. See also on Cabot's involvement Ian S. Evison, "Pragmatism and Idealism in the Professions: The Case of Richard Clarke Cabot, 1868–1939" (PhD diss., University of Chicago, 1995), 275–99.

11. Some diagnoses are difficult to sort out. Did "neurosis of menopause" qualify as a nervous disorder? Only one patient carried the diagnosis, making the distinction less significant.

12. MGH, East Medical Service, 1894, 459:98, 100, 110, PR. See cases of general debility also in MGH, West Medical Service, 1881, 358: 112, 134, PR, Countway.

13. Entry, 8 May 1900, 1:99; entry, 12 Feb. 1900, 1:84; entry, 22 Apr. 1899, 1:51; entry, 12 Jan. 1901, 1:194, PR.

14. Richard C. Cabot, *Differential Diagnosis: Presented Through an Analysis of 385 Cases*, 2d ed. (Philadelphia: W. B. Saunders, 1912), 102–3. All citations from *Differential Diagnosis* in this chapter refer to this widely circulated second edition.

15. Entry, 13 Apr. 1900, 1:93, PR.

16. Entry, 29 July 1899, 1:27; entry, 5 Dec. 1900, 1:186; entry, 8 Mar. 1900, 1:199, PR.

17. For a description of these records and the sampling methods, see Christopher W. Crenner, "Introduction of the Blood Pressure Cuff into U.S. Medical Practice: Technology and Skilled Practice," *AIM*, 1998, 128 (6): 488–93.

18. Robert A. Aronowitz, "When Do Symptoms Become a Disease?" *AIM*, 2001, 134 (9 pt. 2): 803–8.

19. Entries at 8:40, 23:126, 8:116, PR.

20. MGH, East Medical Service, 1909, 743:69, PR, Countway.

21. The first survey of the issues appeared in Hale, *Freud and the Americans*. A detailed analysis of the activity of the Psychopathic Hospital appears in Elizabeth Lunbeck, *The Psychiatric Persuasion: Knowledge, Gender, and Power in Modern Amer-*

NOTES TO PAGES 149–154   281

ica (Princeton, N.J.: Princeton University Press, 1994). The career of Isador Coriat also shed light on this development in Boston: Barbara Sicherman, "Isador H. Coriat: The Making of an American Psychoanalyst," in *Psychoanalysis, Psychotherapy, and the New England Medical Scene, 1894–1944,* ed. George E. Gifford (New York: Science History Publications, 1978), 163–80.

22. Richard C. Cabot, *A Guide to the Clinical Examination of the Blood for Diagnostic Purposes* (New York: William Wood, 1896), 273.

23. Entries, 12 and 26 Jan. 1901, 1:194, PR. See, however, a later use in a similar manner, after Cabot's denunciation of the placebo, in entry, 24 June 1912, 23:13, PR.

24. A more general argument about the constraints on practice in the late nineteenth century appears in John Harley Warner, *The Therapeutic Perspective: Medical Practice, Knowledge, and Identity in America, 1820–1885* (Cambridge, Mass.: Harvard University Press, 1986).

25. One exception was a prescription for Blaud's pill given for psychoneurosis: entry, 24 June 1912, PR, 23:13. Recall, however, that this medication was the "placebo" in the treatment of debility in 1900, 1:194, PR.

26. 15:47, 72, 18, PR; 16:20, PR; 11:119, PR.

27. Entries, 8 May 1900, 1:100, PR; 9 May 1900, 1: 101, PR.

28. For a contemporary example of work therapy for nervous conditions, see W. B. Hopkins, "The Doctor: What Are the Requirements," *Wisconsin Medical Journal,* 1913, 12 (1): 21–25.

29. Entry, 31 May 1912, and letter to RCC, 11 Oct. 1912, 22:132, PR; and letter to RCC, 5 Oct. [?1914], 28:122, PR.

30. Ibid.

31. On the rise of neurasthenia in the United State, see esp. F. G. Gosling, *Before Freud: Neurasthenia and the American Medical Community, 1870–1910* (Urbana: University of Illinois Press, 1987), who also suggests that the success of neurasthenia created the basis for its decline. To compare a European case, see the discussion of the rise of neurasthenia in France in Christopher E. Forth, "Neurasthenia and Manhood in Fin-de-Siècle France," in *Cultures of Neurasthenia from Beard to the First World War,* ed. Marijke Gijswijt-Hofstra and Roy Porter (New York: Rodopi, 2001), 329–61.

32. New England Baptist Hospital, *Fifth Annual Report of the Superintendent* (Boston: New England Baptist Hospital, 1898), 34.

33. St. Vincent's Hospital, *Fifteenth Annual Report of St. Vincent's Hospital* (Worcester, Mass.: Oliver Wood, 1909) 10 ff.

34. A. S. Hershfield, "Neurasthenia," *Illinois Medical Journal,* 1916, 29 (5): 341–44, quotation from 344.

35. Frank G. Murphy, "The Psycho-Neuroses and the General Practitioner," *Journal of the Iowa State Medical Society,* 1912, 2 (2): 558–563, quotations from 558, 560–61.

36. G[eorge] L[incoln] Walton, "Distinction Between the Psychoneuroses Not

Always Necessary," *BMSJ*, 1909, 161:471–74. Other neurologists disagreed. Philip Coombs, "The Rehabilitation of Neurasthenia," *BMSJ*, 1910, 162:269–73, argued, for example, that neurasthenia represented "a real nervous exhaustion . . . which demands our best therapeutic efforts on the physical side."

37. Letter to RCC, 5 Nov. 1909, 14:100, PR.

38. Cabot, *Differential Diagnosis*, 114.

39. RCC, letter, 5 Oct. 1912; and letter to RCC, n.d., 23:40, PR.

40. Letter to RCC, n.d., 23:40 PR.

41. RCC, letter to Dr. Durant, n.d [1916], 22:86, PR.

42. Letter to RCC, 2 Aug. 1920, 35:91, PR.

43. Letter to RCC, 12 Jan. 1912, 16:40, PR.

44. Letter to RCC, n.d., 7:222, PR.

45. Letter to RCC, 3 Feb. 1914, 26:146, PR.

46. MGH, East Medical Service, 1906, 651:737; 1906, 655:35, PR.

47. A provocative analysis of the issue of countertransference in the psychoanalysis of hysteria may be read in Lucien Israël, *L'Hystérique, le sexe et le médecin* (Paris: Masson, 1979).

48. MGH, East Medical Service, 1910, 27:763, PR; MGH, West Medical Service, 1895, 476:47, PR, Countway.

49. J. N. Hall, letter to RCC, 31 Dec. 1906, 6:241, PR.

50. MGH, East Medical Service, 763:27, 673:137, PR.

51. MGH, East Medical Service, 1909, 743:175; 465:75, PR.

52. Richard C. Cabot, *Social Work: Essays on the Meeting Ground of Doctor and Social Worker* (Boston: Houghton Mifflin, 1919), 98.

53. For the broad cultural relevance of syphilophobia in this period and slightly later, see Allan M. Brandt, *No Magic Bullet: A Social History of Venereal Disease in the United States since 1880,* expanded ed. (New York: Oxford University Press, 1987), 50, 155–60.

54. Entry, 12 Apr. 1902, and letter to RCC, June, 1908, 2:60, PR.

55. We might speculate whether Cabot's reading of Freud led him to wonder if a pipe is always just a pipe.

56. Entry, 30 Oct. 1900, 1:160, PR; for the contemporary use of an outside laboratory to examine the sputum for tubercle, see 1:155, PR.

57. For a chronology of changing American debates on race, see John Higham, *Strangers in the Land: Patterns of American Nativism, 1860–1925,* new ed. (New Brunswick: Rutgers University Press, 2002); and on the rise and impact of scientific views on race: ibid., 149–57. Also see for the decline of these views in the face of persisting anti-Semitism: Elazar Barkan, *The Retreat of Scientific Racism: Changing Concepts of Race in Britain and the United States between the World Wars* (New York: Cambridge University Press, 1992), 210–20.

58. For examples in Cabot's practice, see 11:126; 8:116; 8:178; 5:42; and 18:20, PR. And see a case diagnosed as "Hebraic neurasthenia" in MGH, East Medical Service, 1907, 673:137, PR. Cabot was not on service at the hospital at that time. The

attending physicians were Drs. James Minot and William Gannett. See also the comprehensive report on Jewish nervousness at the hospital in H. Morrison, "A Study of Fifty-One Cases of Debility in Jewish Patients," *BMSJ*, 1907, 157 (25): 816–19. For a case study of the medical reasoning about Jewishness and tuberculosis, see Alan Kraut, *Silent Travelers: Germs, Genes, and the "Immigrant Menace"* (New York: Basic Books, 1994), 136–65. For anti-Semitic influence on psychiatric diagnosis in the context of French psychoanalysis, see Jan Goldstein, "The Wandering Jew and the Problem of Psychiatric Anti-Semitism in Fin-de-Siècle France," *Journal of Contemporary History*, 1985, 20:521–52. For a theoretical understanding of the issue, I have relied especially on the introduction to Sander Gilman, *Freud, Race, and Gender* (Princeton, N.J.: Princeton University Press, 1993). A helpful survey of the medical literature on Jewish nervousness appears in Edward Shorter, *From Paralysis to Fatigue: A History of Psychosomatic Illness in the Modern Era* (New York: Free Press, 1992) 90–117.

59. Arthur J. Linenthal, *First a Dream: The History of Boston's Jewish Hospitals, 1896 to 1928* (Boston: Beth Israel Hospital and Countway Medical Library, 1990).

60. Morrison, "Study of Fifty-One Cases of Debility in Jewish Patients," 818, 817.

61. Maurice Fishberg, *The Jews: A Study of Race and Environment* (New York: Scribner, 1911), vii. In his "Study of Fifty-One Cases of Debility in Jewish Patients," Morrison echoed a larger debate among Jewish social scientists and physicians about the influence of immigration on the character of the Jewish people. Some opponents of assimilation within the Zionist movement in Europe argued that detrimental effects of immigration such as the putative increase in nervous disorders that did lasting, fundamental damage indicated the need for a separate homeland. More moderate assimilationists, like Dr. Maurice Fishberg, cited by Morrison, proposed that these negative effects of immigration were transient and remediable with improving conditions of life. See an elegant analysis of this debate in Mitchell B. Hart, "Racial Science, Social Science, and the Politics of Jewish Assimilation," in *Science, Race, and Ethnicity: Readings from Isis and Osiris*, ed. John P. Jackson Jr. (Chicago: University of Chicago Press, 2002), 99–128.

62. Morrison, "Study of Fifty-One Cases of Debility in Jewish Patients," 818, 817.

63. Hart, "Racial Science," illustrates cogently this heterogeneity of political emphasis among scholars debating putative racial disparities in the health of the Jewish people.

64. See very similar generalizations in Osler, *Principles and Practice*, 185, on tuberculosis and racial incidence for example.

65. Cabot, *Differential Diagnosis*, 382, 141, 173, 384.

66. Ibid., 213, 158.

67. Editorial, "Diabetes Mellitus," *JAMA*, 1901, 36 (9): 574–75. For an examination of the extensive literature on Jewish predisposition to diabetes, see esp.

Gilman, *Freud, Race, and Gender*, 129. Gilman cites, as an analogous example, a letter from William Osler describing the influence of nervous excitement on higher rates of diabetes among Jews.

68. Bernheim's letter was a response to an editorial attributing the differences in the expression of diabetes among the Jews to differences in the blood: Editorial, *JAMA*, "Diabetes Mellitus," 574–75. He rejected this hypothesis: Albert Bernheim, "Correspondence," *JAMA*, 1901, 36 (12): 825.

69. The differences in bodily disease often seemed to medical analysts to stem from cultural practices and habits: Gilman, *Freud, Race, and Gender*, 62–63, 128–29, 170–72, 179–80. A comparable examination of theories on putative Jewish variability in the expression of bodily disease appears in Sander Gilman, *The Jew's Body* (New York: Routledge, 1991), 98–200.

70. Editorial, "Paresis Among Jews," *JAMA*, 1900, 35 (20): 1285.

71. George M. Frederickson, *Racism: A Short History* (Princeton, N.J.: Princeton University Press, 2002).

72. The similarities of disease are noted in E. M. Hummel, "The Rarity of Tabetic and Paretic Conditions in the Negro," *JAMA*, 1911, 56 (22): 1645–46. See comparable explanation in Thomas W. Murrell, "Syphilis and the American Negro," *JAMA*, 1910, 54 (11): 846–49. Cf. also the similar use of "mental predisposition" to explain the differential effects of tuberculosis in different racial groups: C. R. Grandy, "The Negro Consumptive [Abstract]," *JAMA*, 1908, 51 (6): 533.

73. S. T. Barnett, "Appendicitis in the Negro," *JAMA*, 1908, 51 (24): 2081; P. Gourdin, "Obstetrics among the Negroes of South Carolina," *American Journal of Obstetrics*, 1893, 28 (5): 717–18.

74. Note, RCC to patient, included with 4:25, PR. On cooperation with a lay mental therapist, for example, see the entries and correspondence at 7:27, and 28:13, PR.

75. The quotation from the patient with psychoneurosis is at 22:132, PR. For a similarly extended series of notes on the care of a nervous patient who discussed with Cabot some details of her marriage and sexual life, see entries 7:237, 7:255, 8:60, PR.

76. Virginia A. M. Quiroga, *Occupational Therapy: The First 30 Years, 1900 to 1930* (American Occupational Therapy Association, 1995), 93–111. "Obituary: Herbert James Hall," *JAMA*, 1923, 80 (10): 713.

77. Herbert J. Hall and Mertice M. C. Buck, *The Work of Our Hands: A Study of Occupations for Invalids* (New York: Moffat, Yard & Co, 1915), 33–45, 57–60.

78. Ibid., 57, 59. See also Herbert J. Hall, "The Systematic Use of Work as a Remedy in Neurasthenia and Allied Conditions," *BMSJ*, 1905, 152:30–32.

79. Hall and Buck, *Work of Our Hands*, 35, 38.

80. Cabot, *Differential Diagnosis*, 223; and the record entry at 45:21, PR.

81. For a survey of the public issues of American anti-Semitism in the twentieth century see Arthur A. Goren, *The Politics and Public Culture of American Jews* (Bloomington: Indiana University Press, 1999). A careful examination of the his-

torical process of Jewish assimilation in the setting of changing racial theories is found in Matthew Frye Jacobson, *Whiteness of a Different Color: European Immigrants and the Alchemy of Race* (Cambridge, Mass.: Harvard University, 1998), esp. 171–99.

82. MGH, East Medical Service, 673:137, PR. Cabot, *Differential Diagnosis*, 190–91.

83. On the tendentious relationship between MGH and the Jewish residents of the West End, see Linenthal, *First a Dream*, 20, which includes the quotation on lying. On the social hierarchies of early twentieth-century Boston, see Frederic Cople Jaher, *The Urban Establishment: Upper Strata in Boston, New York, Charleston, Chicago, and Los Angeles* (Urbana: University of Illinois Press, 1982). On late nineteenth-century Jewish immigration to Boston, see Sam Bass Warner Jr., *Streetcar Suburbs: The Process of Growth in Boston, 1870–1900*, 2d ed. (Cambridge, Mass.: Harvard University Press, 1978) 11, 79, 80, 97. On the anti-immigration politics of the period and Senator Henry Cabot Lodge, see Desmond King, *Making Americans: Immigration, Race, and the Origins of the Diverse Democracy* (Cambridge, Mass.: Harvard University Press, 2000), 50–81. See also Jacobson, *Whiteness of a Different Color*, 182–83.

84. Cabot, *Differential Diagnosis*, 551–52.

85. Ibid., 265.

86. Ibid., 108, 117.

87. Richard C. Cabot, *Foregrounds and Backgrounds in Work for the Sick* (Boston: New England Hospital for Women and Children, 1906), unpaginated [1].

88. Linenthal, *First a Dream*, 372–74, 37, 40, 638, n. 96.

89. Ibid., 516 n. 3.

90. Letter to RCC, 15 Dec. 1910, 6:18, PR.

91. Entry, 16 Dec. 1910; letter to RCC, 15 Dec. 1910; and letter to RCC, 2 Dec. 1913, 18:6, PR.

92. Entry, 10 Jan. 1908; letter to RCC, 29 Jan. 1909, 9:19, PR. Cabot, *Differential Diagnosis*, 32.

93. Letter to RCC, 29 Jan. 1909, 9:19, PR.

94. Entries and correspondence at 7:237, PR. Cabot described this patient's difficulties in an earlier entry in this way: "Her suffering isn't pain but a belt of extreme discomfort c[um] mental diffusion so that she can't hold herself together or speak" (entry, 29 Apr. 1907, 7:100, PR).

CHAPTER 6. MEDICAL CARE FOR THE DYING, IN PRINCIPLE AND IN FACT

1. "Notes of APM Discussion with RCC 1936," box 18, Research Materials of Ada P. McCormick, Cabot Papers.

2. J. Elliot Cabot, letter to RCC, 16 Apr. 1893, fld. "J Elliot Cabot," box 2, Correspondence with Family and Friends, Cabot Papers.

3. Notebook, "Edward T. Cabot Eq. Jur," fld. "Diary 1932," box 3, Biographical Material, Cabot Papers.

4. RCC, letter to Mea Coolidge, n.d.; untitled note, n.d.; RCC, letter to Mea Coolidge, 1 Feb. 1892; RCC, letter to Dearest Mother, 19 June 1900, all in fld. "Letters to Various Correspondents," box 16, Correspondence, Cabot Papers.

5. RCC, letters to "Dearest Mother," 26 June, 5 July, 19 July, 5 Aug., fld. "Elizabeth Cabot," box 2, Correspondence with Family and Friends, Cabot Papers. For a brief review of the conventional care of diabetes in this period, see Chris Feudtner, *Bittersweet: Diabetes, Insulin, and the Transformation of Illness* (Chapel Hill: University of North Carolina Press, 2003), 44–49.

6. For cases of diabetes from the time at this hospital, see MGH, West Medical Service, 1894, 460:8, 184, PR, Countway; RCC, letter to "Dearest Mother," 5 Aug., fld. "Elizabeth Cabot," box 2, Correspondence with Family and Friends. On the nineteenth-century management of diabetes, see also Chris Feudtner, "The Want of Control: Ideas, Innovations, and Ideals in the Modern Management of Diabetes Mellitus," *BHM*, 1995, 69 (Spring): 66–90.

7. RCC, letter to Ella Lyman, 4 Nov. 1893, box 4, Research Material of Ada P. McCormick, Cabot Papers.

8. "Interview with Hugh Cabot, November 1938," box 19, Research Material of Ada P. McCormick, Cabot Papers.

9. Interview, fld. "Dialogues with RCC. Ted's Death," box 18, Research Materials of Ada P. McCormick, Cabot Papers.

10. Ibid.

11. "Interview with Hugh Cabot," Research Materials of Ada P. McCormick, fld. "Interviews," box 19, Cabot Papers.

12. W. Bruce Fye, "Active Euthanasia: An Historical Survey of Its Conceptual Origins and Introduction into Medical Thought," *BHM*, 1978, 52 (4): 492–502. Stressing the positive support for variants of the practice among physicians is Russell Hallander, "Euthanasia and Mental Retardation: Suggesting the Unthinkable," *Mental Retardation*, 1989, 27 (2): 53–61. Stressing the opposition of American physicians is Ezekiel J. Emmanuel, "The History of Euthanasia Debates in the United States and Britain," *AIM*, 1994, 121 (10): 793–802. A synoptic review appears also in Robert P. Hudson, "The Many Faces of Euthanasia," *Medical Heritage*, 1986, 2 (2): 102–7.

13. Fye, "Active Euthanasia," 492–502.

14. Dictionaries with no entry for the word "euthanasia" include Bartholomew Parr, *London Medical Dictionary* (Philadelphia: Mitchell, Ames & White, 1819); G. Motherby, *A New Medical Dictionary* (London: J. Johnson, 1775); R. James, *A Medicinal Dictionary* (London: T. Osbourne, 1745); Robert Hooper, *Lexicon Medicum or Medical Dictionary*, 6th ed. (London: Longman, Rees, Orme, Brown, & Green, 1831); and Richard D. Hoblyn, *A Dictionary of Terms Used in Medicine* (London: Sherwood, Gilbert, & Piper, 1835). The term was discussed in the medical literature of the day, however; see, e.g.: Nicolai Paradys, *Oratio de ευθανασία naturali et quid ad eam conciliandam medicina valeat* (Lyden: Henricum Mostert, 1794), cited also in

abbreviated form by William Munk, *Euthanasia, or, Medical Treatment in Aid of an Easy Death* (London: Longmans, Green, 1887), [v].

15. On the classical meaning of *euthanasia,* see M. Tullius Cicero, *Ad Atticum* 16.7.3, in *Letters to Atticus,* ed. L. C. Purser, where it signifies a good or noble death (available as searchable text through the Perseus Digital Library at www.perseus .tufts.edu/ [accessed spring 2004]). For the nineteenth century, see Chapin Harris, *Dictionary of Medical Terminology,* 2d ed. (Philadelphia: Lindsay & Blakiston, 1855), 273, which defines "euthanasia" as "an easy death." See also George M. Gould, *An Illustrated Dictionary of Medicine Biology and Allied Sciences,* 3d ed. (Philadelphia: P. Blakiston & Son, 1897), and id., *A Pocket Medical Dictionary* (Philadelphia: P. Blakiston's Son, 1900). Richard Quain, *A Dictionary of Medicine* (New York: D. Appleton, 1883), has no entry for the word.

16. George M. Gould, *Gould's Medical Dictionary,* 2d ed. (Philadelphia: P. Blakiston's Son, [1926]) gives the second definition as "the killing of people who are suffering from an incurable or painful disease." Only the easy death definition appears in W. A. Newman Dorland's *American Illustrated Medical Dictionary,* 18th ed. (Philadelphia: W. B. Saunders, 1940), but starting with the 20th edition (Philadelphia: W. B. Saunders, 1945), it also defines the term as meaning "mercy death; the putting to death of a person suffering from an incurable disease," a definition more revealing of tensions than the average dictionary entry.

17. A. S. Duncan, G. R. Dunstan, and R. B. Welbourn, eds. *Dictionary of Medical Ethics* (London: Darton, Longman, & Todd, 1977).

18. "Code of Ethics of the American Medical Association," *Transactions of the American Medical Association,* 1857, 10:607–20, included in all the early volumes of the transactions. Discussed briefly in Peter G. Filene, *In the Arms of Others: A Cultural History of the Right-to-Die in America* (Chicago: Ivan R. Dee, 1998), 3. Similar advice from a slightly later period under a pragmatic guise can be found in Daniel W. Cathell, *Book on the Physician Himself, and Things that Concern His Reputation and Success* (Philadelphia: Davis, 1892), 31–41.

19. W. R. Albury, "Ideas of Life and Death," in *Companion Encyclopedia of the History of Medicine,* ed. W. F. Bynum and Roy Porter (London: Routledge, 1993), 249–80.

20. William Munk, *Euthanasia, or, Medical Treatment in Aid of an Easy Death* (London: Longmans, Green, 1887).

21. Ibid., 69–71, 74.

22. Alfred Worcester, "The Care of the Dying," in *Physician and Patient: Personal Care,* ed. L. Eugene Emerson (Cambridge, Mass.: Harvard University Press, 1929), 20–24. He later expanded this essay into a book, *The Care of the Aged the Dying and the Dead,* 2d ed. (Springfield, Ill.: C. C. Thomas, 1940); quotations here from 45, 37.

23. Walter B. Forman, Judith A. Kitzes, Robert P. Anderson, et al., *Hospice and Palliative Care: Concepts and Practice,* 2d ed. (Boston: Jones & Bartlett, 2003) esp. a first chapter on the history of palliative care.

24. The quotation is from Worcester, "Care of the Dying," 204. The note on hyoscine is in Cicely Saunders, *Living with Dying: The Management of Terminal Disease* (New York: Oxford University Press, 1983), 32. In the earlier essay Worcester recommended hypodermic injections of atropine for the death rattle, which would be similar in pharmacological effects. Hyoscine is mentioned in the later revised edition of Worcester's essay.

25. The quotation is from Worcester, *Care of the Aged the Dying and the Dead*, 45. Cf. from a slightly later date Cicely Saunders, *Care of the Dying* (London: Macmillan, 1959). A similar approach is evident in John Hinton, *Dying* (London: Penguin Books, 1967).

26. Philippe Ariès, *The Hour of Our Death*, trans. Helen Weaver (New York: Knopf, 1981), 559–601.

27. W. N. Leak, "The Care of the Dying," *The Practitioner*, 1948, 161:80–87, quotation p. 83.

28. *Medizinisches Handlexikon*, 1782, 1:30, quoted in Hans-Joachim von Kondratowitz, "The Medicalization of Old Age: Continuity and Change in Germany from the Late Eighteenth to the Early Twentieth Century," in *Life, Death, and the Elderly: Historical Perspectives*, ed. Margaret Pelling and Richard M. Smith (London: Routledge, 1991), 144. A parallel analysis encompassing a larger range of European sources appears in Albury, "Ideas of Life and Death."

29. C. Mettenheimer, *Nosologische und anatomische Beiträge zu der Lehre von den Greisenkrankheiten* (Leipzig: B. G. Teubner, 1863), 13, as quoted in Kondratowitz, "Medicalization of Old Age," 156.

30. Bayard Holmes, "Surgery of the Kidney," *JAMA*, 1896, 27:533; Francis Watson, "Cases of Abdominal Surgery," *BMSJ*, 1896, 135:193; Wendell Boyd, "Septicemia Following Gonorrhea," *BMSJ*, 1896, 134:214.

31. Nichols, "Report of Cases of Small-Pox," *BMSJ*, 1896, 134:236.

32. Edward Reynolds, "A Case of Spontaneous Rupture of the Uterus," *BMSJ*, 1896, 134:62; John Meachem, "Report of Fifteen Cases of Puerperal Eclampsia," *JAMA*, 1890, 15:275; H. C. Dalton, "Cases of Penetrating Stab Wounds," *JAMA*, 1890, 15:709. See also cases cited in Peter English, *Shock, Physiological Surgery, and George Washington Crile: Medical Innovation in the Progressive Era* (Westwood, Conn.: Greenwood Press, 1980).

33. Earnest M. Daland, "Palliative Treatment of the Patient with Advanced Cancer," *JAMA*, 1948, 136 (6): 391–96, quotations from 391, 392.

34. Albury, "Ideas of Life and Death," 268–70.

35. The quotation appears in Renée C. Fox and Judith P. Swazey, *The Courage to Fail: A Social View of Organ Transplantation and Dialysis* (Chicago: University of Chicago Press, 1974), 113–15, 319–25, 323. Albury, "Ideas of Life and Death," traces the roots of this concept. See also the discussion in Nicholas A. Christakis, *Death Foretold: Prophecy and Prognosis in Medical Care* (Chicago: University of Chicago Press, 1999).

36. Worcester, *Care of the Aged the Dying and the Dead*, 47.

37. Cicely Saunders, "Appropriate Treatment, Appropriate Death," in *The Management of Terminal Disease,* ed. id. (London: Arnold, 1978). 1–9. The quotation is from Saunders, *Living with Dying,* 5.

38. William F. Mengert, "Terminal Care," *Illinois Medical Journal,* 1957, 112 (3): 99–104.

39. RCC, letter to Dr Wheeler, 29 Feb. 1914, 27:10, PR.

40. RCC, letter to family member, 18 Feb. 1914, 27:9, PR.

41. Letter to RCC, 30 July 1915, 6:63, PR.

42. Munk, *Euthanasia,* 73, 76; Worcester, "Care of the Dying," 211–12; Derek Doyle, Geoffrey W. C. Hanks, and Neil MacDonald, eds., *Oxford Textbook of Palliative Medicine* (New York: Oxford University Press, 1994), 501–2. On the promulgation of the idea of medical euthanasia, see Fye, "Active Euthanasia."

43. Entry, Sept. 1900, 1: 150, PR.

44. Ibid.

45. Dr. Perley, letter to RCC, 9 Nov. 1913, 25:142, PR.

46. Letter to RCC, 8 June 1908, 2:162, PR.

47. Dr. Smith, letter to RCC, 27 Feb. 1914, 27:13, PR.

48. Letter to RCC, 18 Feb. 1915, 22:6, PR; Dr. Smith, letter to RCC, 27 Feb. 1914, 27:13, PR.

49. These records are filed with the "Biographical Materials" in the Cabot Papers and not with the separate restricted collection of patient records. I have accordingly treated them as part of a collection intended to be used for future scholarship without special restrictions.

50. "Notes on R.C.C.'s cardiac weakness," 19 Jan. 1939, fld. "Notes on Dr. Cabot's illness," box 1, Biographical Materials, Cabot Papers

51. Ibid. See also the discussion of heart failure in Richard C. Cabot, *Facts on the Heart* (Philadelphia: W. B. Saunders, 1926), 770.

52. "Ella's Last Summer," fld. "Dialogues with RCC," box 18, Research Materials of Ada McCormick, Cabot Papers.

53. Ibid.

54. Ibid. Also see "Interview with Philip Cabot," 2 Nov. 1936, fld. "Interview with Philip Cabot," box 19, Research Materials of Ada McCormick, Cabot Papers.

55. "Ella's Last Summer," 2.

56. Ibid., and "Interview with Philip Cabot," Cabot Papers.

57. "Ella's Last Summer," 2.

58. Ibid.

CHAPTER 7. FROM CABOT'S DAY TO OURS

1. For an example of an early attack from within the medical profession on the doctrine of informed consent in medical decision-making see: N. J. Demy, "Informed Opinion on Informed Consent," *JAMA,* 1971, 217 (5): 696–97.

2. David J. Rothman, *Strangers at the Bedside: A History of How Law and*

Bioethics Transformed Medical Decision Making (New York: Basic Books, 1991). Rothman's account is the fullest historical version of the argument that there has been a transformation of the negotiations between doctors and patients under external pressure. A detailed coverage of similar issues related to clinical management of breast cancer appears in Barron H. Lerner, *The Breast Cancer Wars: Hope, Fear and the Pursuit of Care in Twentieth-Century America* (New York: Oxford University Press, 2001), esp. 141–69. Ruth Faden and Tom Beauchamp in *A History and Theory of Informed Consent* (New York: Oxford University Press, 1986), 86–101, review the historical emergence of informed consent briefly but with similar conclusions. A parallel, concise version of the case of informed consent can be found also in John Fletcher, "The Evolution of the Ethics of Informed Consent," in *Research Ethics*, ed. Kåre Berg, and Knut Tranøy (New York: Alan R. Liss, 1983), 187–228. For a diverging view on the question of change in the nature of patient-doctor relationships, see Jay Katz, *The Silent World of Doctor and Patient* (New York: Free Press, 1984). Katz and Rothman both endorse the progressive value of greater individual power for patients, but Katz denies that much progress has been achieved. For a helpful overview of the past few decades of fundamental change in the American medical system, see David Dranove, *The Economic Evolution of American Health Care: From Marcus Welby to Managed Care* (Princeton, N.J.: Princeton University Press, 2000).

3. See Fletcher, "Informed Consent," 217. A historically based version of a similar appreciation for Cabot appears in Chester Burns, "Richard Clarke Cabot (1868–1939) and Reformation in American Medical Ethics," *BHM*, 1977, 51: 353–68, esp. n. 49. Laurie O'Brien understands Cabot in light of contemporary ethical debates more as a conservative critic of secular professionalization in medicine: " 'A Bold Plunge into the Sea of Values': The Career of Dr. Richard Cabot," *New England Quarterly*, 1985, 58 (4): 533–53.

4. A wonderfully detailed statement of how paternalism might ideally function appeared later in the sociologist Talcott Parsons's model of the sick role from the 1950s: see Parsons, "Illness and the Role of the Physician: A Sociological Perspective," *American Journal of Psychiatry*, 1951, 21: 452–60; and "The Sick Role and the Role of the Physician Reconsidered," *Milbank Memorial Fund Quarterly*, Summer 1975, 257–78. Also see the chapter on the physician in id., *The Social System* (Glencoe, Ill.: Free Press, 1951).

5. On informed consent, see Faden, *Informed Consent*, 100. Faden and Beauchamp are appropriately circumspect about the degree to which practice has changed in medicine, in line with the changes in ideals. Also see B. R. Cassileth et al., "Informed Consent—Why Are Its Goals Imperfectly Realized," *NEJM*, 1980, 302:896–900.

6. The concern about the boundaries of medical ethics and its relationship to medical practice is not esoteric to the enterprise of bioethics. Cf. Mark Siegler, "Medical Ethics as a Medical Matter," and Robert Veatch, "Who Should Control the Scope and Nature of Medical Ethics," in *The American Medical Ethics Revolution*, ed.

Robert Baker, Arthur Caplan, Linda Emanuel, and Stephen Latham (Baltimore: Johns Hopkins University Press, 1999), 158–79. A view from sociology that is sympathetic to the problem appears in Renée Fox, "The Evolution of American Bioethics: A Sociological Perspective," in *Social Science Perspectives on Medical Ethics,* ed. George Weisz (Philadelphia: University of Pennsylvania, 1990), 201–17. A more skeptical view can be found in Robert Martensen, "The History of Bioethics: An Essay Review," *Journal of the History of Medicine and Allied Sciences,* 2001, 56 (2): 168–75.

7. Caroline Kaufman has identified a fascinating piece of evidence for this pattern in the appearance of the discussion of informed consent in the medical literature earlier than in the legal literature: "Informed Consent and Patient Decision Making: Two Decades of Research," *Social Science and Medicine,* 1983, 17 (21): 1657–64.

8. The earlier survey is in Cassileth, "Informed Consent." Cf. K. C. Saw et al., "Informed Consent: An Evaluation of Patients' Understanding and Opinion," *Journal of the Royal Society of Medicine,* 1994, 87:143–44.

9. Mark R. Chassin and Elise C. Becher, "The Wrong Patient," *AIM,* 2002, 136 (11): 826–33.

10. Ibid., 31.

11. Carl Schneider works out a full analysis of informed consent as consumer choice in *The Practice of Autonomy: Patients, Doctors, and Medical Decisions* (New York: Oxford University Press, 1998), esp. 186–227. His analysis of informed consent develops in detail many of the ideas that I hit upon independently when I started this chapter. I am grateful to an anonymous reviewer for the Johns Hopkins University Press for bringing this book to my attention after an earlier draft. I differ from Schneider, however, in my understanding of the relationship between bureaucratic organization and patient decision-making. See also the review of this book by David Rothman, "Revisionism Misplaced: Why This is Not the Time to Bury Autonomy," *Michigan Law Review,* 1999, 97 (6): 1512–17.

12. See Chapter 3 above.

13. Donald Oken, "What to Tell Cancer Patients," *JAMA,* 1961, 175 (13): 1120–28. Cf. also William T. Fitts Jr., "What Philadelphia Physicians Tell Patients with Cancer," *JAMA,* 1953, 153 (10): 901–4.

14. D. H. Novack, "Changes in Physicians' Attitudes Toward Telling the Cancer Patient," *JAMA,* 1979, 241:897–900.

15. Fitts, "What Philadelphia Physicians Tell Patients with Cancer," 903: "A responsible member of the family should always be told." Cf. also Oken, "What to Tell Cancer Patients," 1123: "Agreement was essentially unanimous that some family member must be informed if the patient is not made aware of the diagnosis."

16. Current regulation seems now to restrict the ability of health-care providers to tell a family member a patient's medical information without the approval of the patient. A parallel development in the 1970s confirms this dynamic of neglecting the prior involvement of the family in the critique of paternalism.

Successful efforts by women advocates in the 1970s gave patients greater control in deciding about the surgical treatment of breast cancer. Reformers of the day justly attacked the paternalistic power of male physicians in excluding women patients from decisions about surgery. Theirs was a satisfying success. But the existing practice that reform overcame, while equally egregious, was different from what opponents portrayed. Surgeons had commonly been seeking consent from a woman's husband. See a mention of this change in Lerner, *Breast Cancer Wars*, 175.

17. Emergency medicine as a specialized practice still does not maintain exclusive control over the practice of medicine in emergency rooms, however: R. Krome, "Twenty-Five Years of Evolution and Revolution: How the Specialty Has Changed," *Annals of Emergency Medicine*, 1997, 30 (5): 689–90.

18. R. M. Wachter, "The Evolution of the Hospitalist Model in the United States," *Medical Clinics of North America*, 2002, 86 (4): 687–706. One has to be cautious of the powerful pull of nostalgia in many of these discussions, however. As early as the 1920s, Francis Peabody lamented the fragmentation of medical care and the loss of a close personal connection to a single physician: Francis W. Peabody, *Doctor and Patient* (New York: Macmillan, 1930), 45–50.

19. Schneider, ibid., 195–210. See an excellent introduction to the issue of client control and consumer empowerment related to medicine in Nancy Tomes, "Merchants of Health: Medicine and Consumer Culture in the United States, 1900–1940," *Journal of American History*, September 2001, 519–47.

20. Eliot Freidson, "Client Control and Medical Practice," *American Journal of Sociology*, 1960, 65: 374–82. See the later development of the idea in Leo G. Reeder, "The Patient-Client as a Consumer: Some Observations on the Changing Professional-Client Relationship," *Journal of Health and Social Behavior*, 1972, 13:406–21. For a clear statement about the inherent resistance to this model among physicians in the period, see Aaron Lazare, Sherman Eisenthal, and Linda Wasserman, "The Customer Approach to Patienthood," *Archives of General Psychiatry*, 1975, 32: 553–58.

21. Letter to RCC, 7 Nov. 1916, 32:106, PR.

22. Letters to RCC, 7 Nov. 1916, and 22 Nov. 1916, 32:106, PR.

23. Robert Burton, *The Anatomy of Melancholy* (1621; New York: Classics of Medicine Library, 1986), 1: 410–12.

24. Carey Woofter, "Dialect Words and Phrases from West-Central West Virginia," *American Speech*, 1927, 2 (May): 347–67. See also Vance Randolph, "A Third Ozark Word-List," *American Speech*, 1929, 5 (October): 16–21.

25. Letter to RCC, 19 May 1902, 2:79, PR.

26. Letter to RCC, 7 May 1902, 1:168, PR.

27. Letter to William Smith, MD, 479 Beacon Street [not hospital address], 19 June 1901, MGH, West Medical Service, 510:128, PR, Countway.

28. Edith Wharton, *Ethan Frome* (New York: Scribner, 1911), 14, 68.

29. *Webster's Imperial Dictionary of the English Language* (Chicago: Geo. W. Ogilvie, 1904) s.v. "doctor."

30. Cf. a very similar use of the term in the dialect used in Ann Sophie Stephens, *High Life in New York* (New York: Bunce, 1854) 175.

31. Sarah Orne Jewett, "An Autumn Holiday," *Harper's New Monthly Magazine*, 1880, 61 (October): 683–91.

32. For this analysis I have relied heavily on the version of gift exchange in James G. Carrier, *Gifts and Commodities: Exchange and Western Capitalism since 1700* (London: Routledge, 1995). The original source for the analysis is Marcel Mauss's "Essai sur le don: Forme et raison de l'échange dans les sociétés primitives," *L'Année sociologique*, 2d ser., 1923–24, translated by I. Cunnison under the title *The Gift: Forms and Functions of Exchange in Archaic Societies* (London: Cohen & West, 1954). Cf. also the valuable critique in Pierre Bourdieu, *Outline of a Theory of Practice*, trans. Richard Nice (London: Cambridge University Press, 1977), 4–9.

33. Carrier, *Gifts and Commodities*, 8–20.

34. Ibid. This analysis follows the work of Carrier closely.

35. I have drawn heavily from a wonderful exposition on the nature of gift relationships relating to blood donation in Richard Morris Titmuss, *The Gift Relationship; from Human Blood to Social Policy* (New York: Vintage Books, 1972). Further exposition of the personal obligations between physicians and patients appears in Eric Cassell, *The Nature of Suffering and the Goals of Medicine* (New York: Oxford University Press, 1991).

36. Interview by Cherie Kelly, 8 Aug. 2000, Lakeview Village, Lenexa, Kansas. Transcript of the interview available from the author on request. Interviewees were kept anonymous as part of an agreement for approval of the research by the Human Subjects Committee, University of Kansas Medical Center.

37. Interview by Christopher Crenner, 17 July 2000, Lakeview Village, Lenexa, Kansas.

38. Interview, 8 Aug. 2000.

39. Interview by Cherie Kelly, 6 July 2000, Lakeview Village, Lenexa, Kansas. This doctor's name has been changed to protect confidentiality under the agreement described in n. 36 above.

INDEX